Wissenschaftliche Reihe Fahrzeugtechnik Universität Stuttgart

Reihe herausgegeben von

André Casal Kulzer, Stuttgart, Deutschland

Hans-Christian Reuss, Stuttgart, Deutschland

Andreas Wagner, Stuttgart, Deutschland

Das Institut für Fahrzeugtechnik Stuttgart (IFS) an der Universität Stuttgart forscht interdisziplinär sowie technologieoffen an modernen und zukunftsorientierten Fahrzeugkonzepten. In enger Zusammenarbeit mit Partnern aus Industrie und Wissenschaft entstehen neue Lösungen für die Mobilität der Zukunft. Das Institut gliedert sich in drei spezialisierte Lehrstühle, die gemeinsam das gesamte Spektrum der Fahrzeugtechnik abdecken: Der **Lehrstuhl für Fahrzeugantriebssysteme** widmet sich der Forschung nachhaltiger Antriebslösungen für künftige Mobilitätskonzepte. Im Fokus stehen alternative, elektrische sowie hybride Antriebssysteme und deren Komponenten – einschließlich der Nutzung nachhaltiger Energieträger wie Wasserstoff, synthetischer Kraftstoffe und Batterien. Der **Lehrstuhl für Kraftfahrzeugmechatronik** beschäftigt sich mit vernetzten, intelligenten und adaptiven Fahrzeugen. Im Zentrum stehen Fragestellungen zum Automatisierten und Vernetzten Fahren, Diagnose, Ladetechnologien, verteilte Systeme sowie softwarebasierte Fahrzeugfunktionen. Der **Lehrstuhl für Kraftfahrwesen** erforscht die physikalischen Grundlagen der Auslegung zukünftiger Fahrzeugkonzepte. Im Mittelpunkt stehen die Bereiche Aerodynamik, Windkanaltechnik, Akustik/NVH, Fahrzeugdynamik, Reifenmanagement und Thermomanagement Gesamtfahrzeug. Das IFS verfügt über eine vielfältige und hochmoderne Forschungsinfrastruktur, die realitätsnahe Untersuchungen vom Einzelbauteil bis zum Gesamtfahrzeug ermöglicht. Besonders hervorzuheben sind der Multikonfigurations- und Antriebsprüfstand für komplexe Antriebskonzepte, der Stuttgarter Fahrsimulator zur Untersuchung menschlichen Fahrverhaltens, der Aeroakustik-Fahrzeugwindkanal für akustische und strömungstechnische Fragestellungen sowie der Thermowindkanal zur Analyse thermischer Prozesse im Gesamtfahrzeug. Die wissenschaftliche Reihe „Fahrzeugtechnik Universität Stuttgart" dokumentiert die im Rahmen von Promotionen am IFS entstandene Beiträge zur Mobilität der Zukunft und zeigt deren thematische Vielfalt, methodische Tiefe und Praxisrelevanz.

Reihe herausgegeben von

Prof. Dr.-Ing. André Casal Kulzer
Lehrstuhl für Fahrzeugantriebssysteme
Institut für Fahrzeugtechnik Stuttgart
Universität Stuttgart
Stuttgart, Deutschland

Prof. Dr.-Ing. Andreas Wagner
Lehrstuhl für Kraftfahrwesen
Institut für Fahrzeugtechnik Stuttgart
Universität Stuttgart
Stuttgart, Deutschland

Prof. Dr.-Ing. Hans-Christian Reuss
Lehrstuhl für
Kraftfahrzeugmechatronik
Institut für Fahrzeugtechnik Stuttgart
Universität Stuttgart
Stuttgart, Deutschland

Marcus Auch

Holistic Modeling Framework for the Investigation of Innovative Battery Thermal Management Concepts

Springer Vieweg

Marcus Auch
Institute of Automotive Engineering
(IFS), Chair of Automotive Engineering
University of Stuttgart
Stuttgart, Germany

Zugl.: Dissertation Universität Stuttgart, 2026 D93

ISSN 2567-0042 ISSN 2567-0352 (electronic)
Wissenschaftliche Reihe Fahrzeugtechnik Universität Stuttgart
ISBN 978-3-658-52611-5 ISBN 978-3-658-52612-2 (eBook)
https://doi.org/10.1007/978-3-658-52612-2

This Springer Vieweg imprint is published by the registered company Springer Fachmedien Wiesbaden GmbH, part of Springer Nature.
The registered company address is: Abraham-Lincoln-Str. 46, 65189 Wiesbaden, Germany

Foreword

This thesis is the result of my time as a research associate and PhD student at the Institute of Automotive Engineering (IFS) at the University of Stuttgart.

First and foremost, I would like to express my sincere gratitude to my doctoral supervisor Prof. Dr.-Ing. Andreas Wagner for giving me the opportunity to conduct the research presented in this thesis at his institute. His guidance and support, as well as his genuine interest in my work, have been invaluable and have significantly shaped my PhD journey.

I would also like to thank Prof. Dr.-Ing. Tjark Siefkes for taking on the responsibility of second reviewer for this thesis.

My thanks extends to my direct supervisor Dr.-Ing. Timo Kuthada for investing his time in numerous discussions about my ideas and methods and especially for his guidance during the early stages of the PhD process.

All my colleagues at IFS and the Research Institute for Automotive Engineering (FKFS) enriched my time not only through their practical, experience-based advice but also by reminding me not to take everything, or myself, too seriously. I would particularly like to thank my university colleagues Dennis Weidner and Marwan Khaled for the insightful discussions and for always being willing to listen. Furthermore, I would like to thank all my students, whose thesis work supported my own PhD research, expanded my knowledge and led to numerous inspiring discussions.

Finally, I am deeply grateful to my parents and my sister for their constant support and for enabling me to pursue my studies. Last but not least, I appreciate the unwavering support and understanding of my partner, especially during late-night work sessions when myself demanded the immediate solution of problems encountered.

Ludwigsburg Marcus Auch

Table of Contents

List of Figures

List of Tables

Nomenclature

Latin Symbols

Symbol	Description	Unit
A	Amplitude	m
A	Surface	m^2
B	Pre-exponential factor	–
b	Width	m
C	Capacitance	F
c_p	Specific heat capacity	$J/kg\,K$
d	Diameter	m
E_a	Activation energy	J/mol
f	Darcy friction factor	–
h	Heat transfer coefficient	$W/m^2\,K$
h	Height	m
I	Current	A
k	Thermal conductivity	$W/m\,K$
l	Length	m
m	Mass	kg
Nu	Nusselt number	–
n	Number	–
P	Power	W
Pr	Prandtl number	–
p	Pressure	Pa
Q	Capacity	Ah

$\dot{Q}$	Heat source	W
$\dot{q}$	Heat flux/Volumetric heat source	$\mathrm{W/m^2}$, $\mathrm{W/m^3}$
R	Universal gas constant	$\mathrm{J/mol\,K}$
R	Resistance	Ω
R_{el}	Electrical resistance	Ω
R_{th}	Thermal resistance	$\mathrm{K/W}$, $\mathrm{K\,m^2/W}$
Ra	Rayleigh number	–
Re	Reynolds number	–
r	Radius	m
T	Temperature	K
t	Time	s
th	Thickness	m
U	Voltage	V
V	Volume	$\mathrm{m^3}$
v	Velocity	$\mathrm{m/s}$
y^*	Dimensionless channel length	–
z	Power law factor	–

Greek Symbols

Symbol	**Description**	**Unit**
α	Dimensionless factor	–
ϵ	Emissivity	–
η	Dynamic viscosity	$\mathrm{N\,s/m^2}$
η	Efficiency	%
λ	Wavelength	m

ρ	Density	$^{kg}/_{m^3}$
σ	Stefan-Boltzmann constant	$^{W}/_{m^2\,K^4}$
τ	Time constant	s

Subscripts

Symbol	**Description**
i	Counting variable
j	Counting variable
0	Ohmic resistance
A	Aging study
act	Actual
amb	Ambient
avg	Average
c	Cell mantle
c	Coulombic efficiency
can	Casing
cond	Conductive
conv	Convective
corr	Correction
cyc	Cyclic
ed	Energy demand
exp	Experimental
f	Fluid
h	Hydraulic diameter
hs	Heat sink

i	Inner
in	Inlet
irr	Irreversible losses
jr	Jelly roll
lo	Local
m	mean
max	Maximum
mix	Mixing losses
N	Nominal
nt	Negative terminal
ntab	Negative tab
o	Outer
p	Polarization resistance
pt	Powertrain
pt	Positive terminal
ptab	Positive tab
rad	Radiative
reac	Side reaction losses
rev	Reversible losses
s	Surface
sim	Simulation
tb	Traction battery
tot	Total
v	Vehicle
w	Wall temperature

Acronyms

0D	Zero-Dimensional
1D	One-Dimensional
3D	Three-Dimensional
BC	Boundary Condition
BEV	Battery Electric Vehicle
BOL	Begin-of-Life
BTMS	Battery Thermal Management System
C	C-rate
CC-CV	Constant-current Constant-voltage
CEI	Cathode Electrolyte Interphase
CFD	Computational Fluid Dynamics
Conv	Conventional BTMS concept
DC	Discharge
ECM	Equivalent Circuit Model
EGL	Ethylene Glycol
EIS	Electrical Impendance Spectroscopy
EOL	End-of-Life
F.S.	Full Scale
FC	Fast Charge
HEV	Hybrid Electric Vehicle
HP	Heat Pipe
HPPC	Hybrid Pulse Power Characterization
HVAC	Heating, Ventilation and Air Conditioning

ICE	Internal Combustion Engine
LFP	Lithium Iron Phosphate
LIB	Lithium-ion Battery
LMO	Lithium Manganese Oxide
ME	Mean Error
MPH-TR	Multi Partition Heat Generation and Thermal Resistance Model
NCA	Lithium Nickel Cobalt Aluminum Oxide
NMC	Lithium Nickel Manganese Cobalt Oxide
OCV	Open Circuit Voltage
p	Parallel
PCM	Phase Change Material
PHEV	Plug-in Hybrid Eletric Vehicle
RC	Resistance-Capacitance
RMSE	Root Mean Squared Error
RMSRE	Root Mean Squared Relative Error
s	Serial
SEI	Solid Electrolyte Interphase
SOC	State of Charge
SOH	State of Health
Split	Split-cooling BTMS concept
Sw	Switchable BTMS concept
VTMS	Vehicle Thermal Management System
WLTC	World Harmonized Light Vehicles Test Procedure

Abstract

Battery electric vehicles (BEVs) have emerged as the most promising solution in the transition toward zero-emission mobility. Meeting the increasing customer expectations for range and charging speed requires efficient battery thermal management systems (BTMSs) capable of removing the waste heat while maintaining a narrow optimal operating temperature range. Since lithium-ion battery (LIB) aging is strongly temperature-dependent, cell-to-cell temperature differences lead to inhomogeneous aging within the BEV LIB module and pack. Therefore, the BTMS must not only regulate the overall operating temperature but also ensure thermal uniformity across all LIB cells.

However, in state-of-the-art unidirectional indirect liquid-cooled BTMS concepts, module-level temperature gradients are inevitable. As the coolant flows through the heat sink and absorbs heat from upstream cells, its temperature rises. As a consequence, downstream cells are operated at higher average temperatures compared to the cells located at the front. This effect results in cumulative temperature-related aging inhomogeneities among the LIB cells over the life-cycle of the BEV.

To address this drawback of indirect liquid cooling approaches, this thesis proposes a switchable BTMS architecture that integrates a 4/2 directional valve to alternate the coolant flow direction between complete load cycles (charging and discharging). The proposed concept maintains an unidirectional flow direction during operation and reverses the flow direction only during vehicle standstill after each complete load cycle. Consequently, the hottest cell in the previous load cycle becomes the coldest in the subsequent cycle and vice versa, effectively reversing the cell-to-cell temperature gradient and homogenizing thermal aging over the BEV life-cycle.

The investigation of the proposed BTMS concept requires a coupled electro-thermal-fluid-aging model. Since this thesis compares the End-of-Life (EOL) characteristics over the entire life-cycle, a 1D modeling approach is necessitated due to the associated computational resources. To still ensure accurate representation of the LIB cell behavior by the 1D model on the full module level, a holistic thermal modeling approach is developed, spanning from individual LIB cell models to a module section and ultimately the full battery module. The modeling approach is applied to a reference battery pack and BTMS design, using a commercially available LIB cell.

At the cell level, coupled 3D and 1D electro-thermal models are developed based on insights gained from a disassembly of the studied LIB cell. These models have been validated at multiple temperature measurement positions and across various load profiles, achieving root mean square errors below 1 K. Furthermore, an Arrhenius-based aging model is parameterized using experimental data, capturing both capacity fade and resistance growth.

At the module section level, the isolated stacked wavy mini-channel heat sink is analyzed using detailed 3D computational fluid dynamics (CFD) simulations, deriving correlations for heat transfer (Nusselt number) and pressure loss (Darcy friction factor). These correlations are applied to the 1D simulations to ensure accurate representation of the heat sink behavior. Furthermore, a representative section of the full module, consisting of 24 LIB cells with applied heat sink cooling, is simulated under a fast-charging scenario using both the developed 3D and 1D models. The 1D model shows good agreement with the more detailed 3D model in predicting the overall jelly roll temperature levels and cell-to-cell temperature differences. Based on this, the 1D module section model is upscaled to represent the full module. With the integration of the derived aging model, the life-cycle characteristics of the investigated BTMS concepts are evaluated.

For the life-cycle study, a load profile consisting of a fast charge and a moderate discharge rate is selected. The results show for the conventional unidirectional BTMS concept, that the LIB cell located at the rear of the module reaches the EOL criterion first, after a total charge throughput of 5100 Ah. In contrast, the proposed switchable BTMS concept increases the charge throughput by 7.2 % before the EOL criterion is met, which directly corresponds to a 7.2 % increase in life-cycle vehicle mileage. An investigation of different switching intervals reveals that the switch interval can be increased up to ten cycles for similar improvements in the charge throughput. Finally, the switchable BTMS concept demonstrates superior performance compared to other advanced concepts, such as the half-split cooling and U-turn heat sink designs.

Kurzfassung

Im Zuge des Wandels der Automobilindustrie hin zur emissionsfreien Mobilität haben sich batterieelektrische Fahrzeuge (BEVs) als vielversprechendste Lösung etabliert. Um den Anforderungen der Kunden an Reichweite und Ladezeiten gerecht zu werden, sind insbesondere effiziente Batterie-Thermomanagementsysteme (BTMS) erforderlich. Da das Alterungsverhalten von Lithium-Ionen-Batteriezellen (LIBs) stark temperaturabhängig ist, führen Zell-zu-Zell-Temperaturunterschiede zu einer inhomogenen Alterung innerhalb eines Batteriemoduls. Daher muss das BTMS nicht nur den engen Betriebstemperaturbereich, sondern auch die Temperaturhomogenität zwischen den einzelnen Zellen zuverlässig sicherstellen.

In aktuellen BEVs wird überwiegend eine indirekte Flüssigkeitskühlung mit einseitiger Durchströmung der Wärmesenke eingesetzt, was zu Temperaturdifferenzen innerhalb eines Batteriemoduls führt. Das Kühlfluid absorbiert die Abwärme der Batteriezellen, wodurch sich dessen Temperatur erhöht. Folglich werden die Zellen stromabwärts der Wärmesenke bei höheren Temperaturen betrieben als die stromaufwärts liegenden Zellen. Dies führt über den Lebenszyklus eines BEVs zu einer kumulativen, temperaturbedingten inhomogenen Alterung der Zellen.

Um diesem Effekt entgegenzuwirken, wird in dieser Arbeit ein umschaltbares BTMS-Konzept untersucht. Dazu wird in die bestehende BTMS-Architektur ein 4/2-Wegeventil integriert. Dieses ermöglicht das Umdrehen der Einströmrichtung des Kühlfluids, was wiederum zu einer Richtungsänderung der Temperaturdifferenz innerhalb des Moduls führt. Das 4/2-Wegeventil wird erst nach einem vollständigen Lade- und Entladezyklus im Fahrzeugstillstand aktiviert. Dadurch bleibt die Strömungsrichtung während des Betriebs durchgehend einheitlich und vermeidet potenzielle Risiken eines Kühlfluidstillstands bei notwendiger hoher Wärmeabfuhr.

Die Bewertung des vorgeschlagenen BTMS-Konzepts fordert ein gekoppeltes elektro-thermisch-strömungsmechanisches Alterungsmodell auf Modulebene. Da die Konzepte über den gesamten Lebenszyklus bewertet werden, muss aufgrund des hohen Rechenaufwands ein 1D-Modellierungsansatz gewählt werden. Um dennoch eine genaue Vorhersage des Zellverhaltens im

1D-Modell auf Modulebene zu ermöglichen, wurde im Rahmen dieser Arbeit ein ganzheitlicher thermischer Modellierungsansatz entwickelt. Dieser umfasst drei Ebenen, von der Einzelzelle über einen Modulausschnitt bis hin zum vollständigen Modul. Für die Anwendung dieses Modellierungsansatzes wurde ein Batteriepaket und BTMS-Konzept auf dem Stand der Technik mit einer kommerziell verfügbaren LIB-Zelle ausgewählt.

Für diesen Ansatz wurden im Rahmen dieser Arbeit ein 3D- und ein 1D-elektro-thermisches sowie ein Arrhenius-basiertes Alterungsmodell entwickelt, parameterisiert und validiert. Zusätzlich wird sowohl im 3D- als auch im 1D-Modell ein Ausschnitt aus dem Gesamtmodul, bestehend aus 24 Zellen, betrachtet. Dabei wird die isolierte Wärmesenke zur Ableitung von Wärmeübergangskorrelationen (Nusselt-Zahl) und Druckverlustkorrelationen (Darcy-Reibungsfaktor) verwendet. Für den Modulausschnitt mit aktiver Kühlung im Schnellladefall konnte eine gute Übereinstimmung zwischen dem 3D- und dem 1D-Modell bei der Vorhersage des Temperaturniveaus sowie der Temperaturunterschiede zwischen den einzelnen Zellen innerhalb des Moduls gezeigt werden. Auf dieser Basis wurde das Modulausschnittsmodell auf die vollständige Modulebene hochskaliert und das entwickelte Alterungsmodell integriert.

Mit diesem Modell wurde die Lebenszyklusanalyse der BTMS-Konzepte durchgeführt. Dafür wurde ein Lastfall definiert, bestehend aus einem Schnellladeprofil und einem moderaten Entladeprofil. Die Ergebnisse zeigen, dass beim konventionellen, einseitig durchströmten BTMS-Konzept die Zellen am Ende des Moduls das Lebensende-Kriterium als Erste nach einem Ladungsdurchsatz von 5100 Ah erreichen, da sie über den gesamten Lebenszyklus den höchsten Temperaturen ausgesetzt sind. Mit dem vorgeschlagenen umschaltbaren BTMS-Konzept hingegen konnte der Ladungsdurchsatz bis zum Lebensende-Kriterium um 7,2 % erhöht werden, was direkt zu einer Laufleistungssteigerung von 7,2 % über den gesamten Lebenszyklus führt. Das Umschaltintervall kann auf bis zu zehn Zyklen erweitert werden, ohne dass die Verbesserungen im Ladedurchsatz signifikant abnehmen. Zusätzlich konnte gezeigt werden, dass das umschaltbare BTMS-Konzept im Vergleich zu einem Split-Kühlungskonzept oder einem U-förmig gestalteten Wärmesenkendesign verbesserte Lebenszykluseigenschaften aufweist.

Nachfolgend sind die wesentlichen Inhalte der einzelnen Kapitel dieser Dissertation nochmals zusammengefasst.

Kapitel 2: Grundlagen und Stand der Technik

Dieses Kapitel führt in die für diese Arbeit relevanten Grundlagen sowie den aktuellen Stand der Technik ein. Zunächst wird ein Überblick über den Energiebedarf von BEVs sowie aktuelle Entwicklungen in Fahrzeug-Thermomanagementsystemen (VTMS) gegeben. Darauf aufbauend wird der Fokus auf die Traktionsbatterie und entsprechende BTMS-Konzepte gelenkt.

In diesem Kontext werden die Funktionsprinzipien von LIB-Zellen, deren gängige Bauformen sowie Zellchemien erläutert, wobei insbesondere der Trend hin zu zylindrischen Formaten hervorgehoben wird. Zur Beschreibung des Zellverhaltens wird anschließend der Stand der Technik in der elektrischen, alterungsbedingten und thermischen Modellierung zylindrischer LIB-Zellen eingeführt. Abschließend werden verschiedene in BTMS-Konzepten eingesetzte Wärmeabfuhrmechanismen diskutiert sowie dic grundlegenden Prinzipien der Wärmeübertragung dargestellt.

Kapitel 3: Forschungsziel

Ziel dieses Kapitels ist es, die bestehende Forschungslücke zu identifizieren und daraus die Forschungsziele dieser Arbeit abzuleiten. Zu diesem Zweck werden zunächst die Einschränkungen und Herausforderungen aktueller indirekter Flüssigkeitskühlkonzepte analysiert. Im Mittelpunkt steht dabei der BTMS-induzierte Anstieg der Betriebstemperatur der LIB-Zellen stromabwärts des Kühlkanals. Dies resultiert über den Lebenszyklus in einer kumulativen, temperaturinduzierten Alterungsinhomogenität zwischen den LIB-Zellen.

Das erste Forschungsziel besteht darin, dieser Problematik durch die Einführung eines umschaltbaren BTMS-Konzepts zu begegnen. Im Gegensatz zu konventionellen, unidirektionalen BTMS-Architekturen ermöglicht dieser Ansatz eine Umkehr der Strömungsrichtung durch die Integration eines 4/2-Wege-Ventils in die bestehende Systemarchitektur. Dadurch kann die Richtung des Temperaturgradienten zwischen den Zellen entlang der Modullänge umgekehrt werden.

Für den quantitativen Vergleich mit dem Stand der Technik wird das zweite Forschungsziel definiert. Hierfür wird ein ganzheitlicher thermischer Modellierungsansatz entwickelt. Dieser folgt einer hierarchischen Struktur, die sich von der Zellebene über einen Modulausschnitt bis hin zum vollständigen Modul erstreckt. Der Ansatz kombiniert Ersatzschaltbildmodelle mit gekoppelten dreidimensionalen und eindimensionalen thermischen Modellen und integriert zudem ein Alterungsmodell auf Zellebene.

Auf Ebene des Modulausschnitts wird insbesondere das Zellverhalten bei aktiver Kühlung untersucht, wobei ein besonderer Fokus auf der detaillierten Modellierung der Wärmeübertragung liegt. Abschließend erfolgt auf Modulebene die Bewertung der BTMS-Konzepte durch Integration des abgeleiteten Alterungsmodells in einen eindimensionalen Modellierungsansatz.

Kapitel 4: Modellierung

Für den Vergleich der BTMS-Konzepte wird in diesem Kapitel ein Referenz-Batteriepackdesign basierend auf dem Stand der Technik eingeführt. Zusätzlich wird eine kommerziell verfügbare LIB-Zelle ausgewählt, um den zuvor eingeführten ganzheitlichen thermischen Modellierungsansatz auf die Referenz-BTMS-Konfiguration anzuwenden.

Auf Zellebene wird ein Ersatzschaltbildmodell (ECM) als elektrisches Modell eingesetzt. Die dafür notwendigen experimentellen Charakterisierungen sowie die Parameteridentifikation werden erläutert. Anschließend wird das ECM mit dreidimensionalen und eindimensionalen thermischen Modellen gekoppelt, die im Rahmen dieser Arbeit entwickelt wurden. Dabei kommt ein Ansatz mit räumlich aufgeteilter Wärmequelle zum Einsatz. Für beide thermischen Modelle werden der Modellaufbau, die thermophysikalischen Eigenschaften sowie die Randbedingungen detailliert beschrieben. Abschließend wird ein Arrhenius-basiertes Alterungsmodell eingeführt.

Auf Ebene des Modulausschnitts umfasst das Referenz-BTMS-Konzept gestapelte, gewellte Mini-Kanal-Wärmesenken. Ein repräsentativer Modulausschnitt bestehend aus 24 Zellen wird modelliert, inklusive des Kühlkanals und des Schaummaterials. Dafür werden sowohl dreidimensionale als auch eindimensionale Modellierungsansätze eingeführt, einschließlich der jeweiligen Modellaufbauten, Materialeigenschaften und Randbedingungen.

Kapitel 5: Validierung

In diesem Kapitel wird die Validierung der entwickelten Zellmodelle sowie des Simulationsansatzes für die Wärmesenke vorgestellt. Hierfür wird die Zelle an drei Messstellen unter verschiedenen Lastfällen untersucht. Die Messstellen befinden sich in der Mitte des Zellmantels sowie an den positiven und negativen Anschlüssen. Als Lastfälle werden repräsentative BEV-Szenarien verwendet, darunter passives Vorheizen, Schnellladen, Laden zu Hause sowie ein dynamischer Entladefall. Die Spannungs- und Temperaturverläufe der LIB-Zelle werden experimentell erfasst und anschließend mit den Simulationsergebnissen des gekoppelten ECM-3D- und 1D-Modells verglichen. Der Vergleich zeigt, dass der mittlere quadratische Fehler (RMSE) der Temperaturvorhersage überwiegend unter 1 K liegt und der Spannungsfehler unter 50 mV bleibt.

Zusätzlich wird eine Alterungsstudie mit fünf der untersuchten LIB-Zellen durchgeführt. Die experimentellen Ergebnisse hinsichtlich Kapazitätsverlust und Anstieg des Innenwiderstands werden genutzt, um die Parameter des semi-empirischen, Arrhenius-basierten Alterungsmodells zu kalibrieren.

Abschließend wird aufgrund fehlender Literaturkorrelationen für gewellte Mini-Kanal-Wärmesenken mit rechteckigem Querschnitt zunächst eine Validierung anhand eines geradlinigen Kanaldesigns durchgeführt. Die aus 3D-Simulationen gewonnenen Wärmeübertragungs- und Druckverlustkennwerte werden mit Literaturdaten verglichen und zeigen eine sehr gute Übereinstimmung. Aufbauend auf diesem validierten Modell wird das gewellte Kanaldesign untersucht. Hierbei werden durch eine parametrische Variation über Reynolds- und Prandtl-Zahlen im BTMS-Betriebsbereich Korrelationen für die Nusselt-Zahl sowie den Darcy-Reibungsfaktor abgeleitet. Diese Korrelationen dienen anschließend als Randbedingungen für den Fluid-Kanal-Wärmeübergang im eindimensionalen thermischen Modell.

Kapitel 6: Ergebnisse

Im Ergebniskapitel wird zunächst die Ebene des Modulausschnitts für ein Schnellladeszenario mit aktiver Kühlung sowohl mit dreidimensionalen als auch mit eindimensionalen Modellierungsansätzen untersucht. Die Ergebnisse der 3D- und 1D-Simulationen zeigen eine sehr gute Übereinstimmung sowohl hinsichtlich der absoluten Temperaturniveaus (RMSE $< 0,5\,K$) als auch der Zell-zu-Zell-Temperaturdifferenzen (0,018 K gegenüber 0,016 K). Diese Resultate belegen, dass das 1D-Modell eine hinreichende Genauigkeit für die Skalierung auf die vollständige Modulebene sowie für weiterführende Analysen von BTMS-Konzepten bietet.

Auf Modulebene werden anschließend Modellreduktionsmethoden angewendet, um die Simulation auf eine einzelne gestapelte, gewellte Mini-Kanal-Wärmesenke sowie die zugehörigen 154 Zellen zu fokussieren. Für die Lebenszyklusanalyse wird ein repräsentatives Lastprofil definiert, bestehend aus einem Schnellladevorgang gefolgt von einer moderaten Entladung. Zusätzlich wird der Betrachtungsrahmen für die Analyse durch die Definition des Lebensende-Kriteriums mit einem State-of-Health (SOH) von 80 % festgelegt. Die Ergebnisse für das konventionelle, unidirektionale BTMS-Konzept werden ausführlich interpretiert und zeigen die erwarteten ausgeprägten temperaturbedingten Alterungsinhomogenitäten über den Lebenszyklus des BEV.

Demgegenüber weisen die Ergebnisse mit dem Einsatz des umschaltbaren BTMS-Konzepts eine Steigerung des gesamten Ladungsdurchsatzes um bis zu 7,2 % auf. Dies entspricht einer Erhöhung der Fahrzeuglaufleistung um 7,2 %. Darüber hinaus werden in diesem Kapitel alternative Umschaltintervalle untersucht. Die Ergebnisse zeigen vergleichbare Vorteile bis zu einem Umschaltintervall von zehn Zyklen. Abschließend wird das umschaltbare BTMS-Konzept mit alternativen fortschrittlichen BTMS-Konzepten verglichen. Hierbei kann eine überlegene Leistungsfähigkeit bei gleichzeitig minimalem Anpassungsbedarf des BTMS-Aufbaus im Vergleich zu Split-Cooling oder U-förmigen Wärmesenkendesigns gezeigt werden.

1 Introduction

In 2015, the Paris Agreement was adopted with the objective of limiting global temperature rise to 1.5 K above pre-industrial levels by the year 2100, in order to mitigate the impacts of climate change [1]. Carbon dioxide, primarily emitted through the combustion of fossil fuels, is recognized as the most significant greenhouse gas driving climate change. Globally, the transportation sector is responsible for approximately one-fifth of total carbon dioxide emissions [2]. While this sector encompasses road, rail, air and maritime transport, road transportation (comprising passenger cars, heavy-duty vehicles and buses) accounts for the largest share, contributing approximately 75 % of transportation-related emissions [2].

To meet the targets set by the Paris Agreement, the European Union has announced a ban on the sale of new internal combustion engine (ICE) vehicles starting in 2035 [3]. As a result, alternative powertrain technologies have been extensively researched to replace conventional ICE systems. While fuel cell vehicles and synthetic fuels are also under investigation, battery electric vehicles (BEVs) have emerged as the most promising and widely adopted powertrain solution for the future of road transportation [4].

As with any technological transition, customers expect the new solution to deliver performance that is comparable to or better than the existing technology. In the context of passenger vehicles, two primary factors influence consumer acceptance, which are cost and comfort. While the cost of BEVs continues to decline and is approaching parity with ICE vehicles, challenges remain in terms of comfort-related performance. In particular, driving range and charging time are critical, as customers expect these aspects to match the convenience and refueling speed of ICE vehicles.

These goals are closely linked to the most critical component in BEVs, the traction battery. At present, lithium-ion batteries (LIBs) are used almost exclusively for this purpose. Their operating principle is based on the reversible conversion of chemical energy into electrical energy and vice versa. Reduced charging times are constrained by two main factors. First, the charging infrastructure must be capable of supplying high currents, which requires adequate energy delivery systems, cooling of charging cables and

M. Auch, *Holistic Modeling Framework for the Investigation of Innovative Battery Thermal Management Concepts*, Wissenschaftliche Reihe Fahrzeugtechnik Universität Stuttgart, https://doi.org/10.1007/978-3-658-52612-2_1

high-power grid connections. Second, the LIB cells themselves must be able to accept high charging currents while maintaining the safety and durability characteristics required for automotive applications.

In efforts to increase the driving range of BEVs, LIB development once again plays a central role. However, the range improvements can also be achieved through the optimization of overall vehicle efficiency across all subsystems, an area that offers a potential unique selling point for BEV manufactures in an increasingly competitive market. As a result, the thermal management system must be emphasized as one of the most critical enablers of system efficiency. The primary function of the vehicle thermal manage- ment system (VTMS) is to ensure optimal operating conditions for four key components, which include the heating, ventilation and air conditioning (HVAC), the power electronics, the electric machine(s) and the traction battery system.

Due to the critical importance of the traction battery system and its relatively narrow optimal operating temperature range of 15 – 35 °C, the battery thermal management system (BTMS) plays a vital role in BEVs [5]. The primary function of the BTMS is to maintain the traction battery within this temperature window, thereby ensuring safety and life-cycle performance requirements. Deviations from the optimal range can significantly accelerate aging, with low temperatures promoting lithium plating, while elevated temperatures enhance the growth of the solid electrolyte interphase (SEI) layer [6, 7, 8]. Consequently, the BTMS must not only regulate temperature within the specified range but also ensure thermal homogeneity, both at the individual cell level and across the battery module or pack.

BTMSs can be categorized based on their heat transfer method into passive, active and hybrid cooling approaches [9]. Among these, active indirect liquid cooling has emerged as the most widely adopted solution for BEVs, owing to its favorable balance between thermal performance, cost and energy consumption [10, 11]. However, one of the key challenges associated with indirect liquid cooling is maintaining temperature homogeneity at the battery module and pack level. In this configuration, a single cooling channel is typically responsible for removing heat from multiple cells. As the coolant flows through the channel and absorbs heat from the cells, its temperature gradually increases along the flow path. This results in a reduced temperature gradient between the coolant and the downstream cells, which leads to the LIB cells located toward the end of the cooling channel operating at higher average temperatures. [12, 13]

To summarize, in state-of-the-art unidirectional indirect liquid cooling BTMS concepts, module-level temperature differences lead to inhomogeneous aging over the life cycle of the BEV. This is identified as a key challenge in BEV design, resulting in two research question to be answered in this thesis.

First, how can BTMS-related aging inhomogeneities within BEV LIB modules be reduced? To address this question, a switchable BTMS concept is proposed in this thesis. Its core principle is the integration of a 4/2-directional valve into the existing BTMS architecture, allowing the coolant-flow direction to alternate after each complete load cycle. As a result, cells that were previously exposed to the highest temperatures become the coldest in the subsequent cycle, and vice versa, thereby reducing cumulative temperature-induced aging inhomogeneities within the module.

Second, how must a reliable modeling workflow be defined to evaluate and investigate BTMS concepts with respect to their impact on the aging behavior of complete BEV battery packs? For this purpose, this thesis defines a holistic thermal modeling framework spanning from the individual cell to the full module level, incorporating the coupled electro-thermal-fluid-aging behavior on the 3D and 1D level.

Consequently, this thesis advances the state of the art by developing and validating a holistic multi-scale electro-thermal-fluid-aging modeling framework for LIB modules and proposing a novel switchable BTMS concept to reduce temperature-induced cumulated aging inhomogeneities. The combination of modeling methodology and concept innovation provides a deeper understanding of BTMS-driven module aging and highlights existing gaps for improved thermal design.

The content of this thesis is structured as follows. In the first chapter, the design principles of BEVs, VTMSs and BTMSs are introduced. This chapter also presents the theoretical background, including the state of the art in LIB cells, their electrical and thermal modeling approaches and the relevant thermodynamic principles governing heat transfer and pressure loss in BTMS heat sinks. Subsequently, the research objective of this work is formulated, covering limitations in the current BTMS state of the art and the novelty of this thesis, consisting of the proposed switchable BTMS concept and the developed holistic thermal modeling approach. The modeling chapter introduces the investigated reference battery pack design and details the modeling approach, progressing from the cell level to the module section level and ultimately to the full module level with both three-dimensional (3D) and one-dimensional (1D) models. Following this, the electro-thermal cell models and aging model are validated, along with the heat sink simulation and the associated heat transfer and pressure loss correlations. With the validation complete, the results at the module section level are analyzed first, with particular emphasis on accurately capturing cell-to-cell temperature differences. Finally, the approach is scaled to the full module level, where the performance of the conventional unidirectional reference BTMS, the proposed switchable BTMS and two additional advanced concepts (split cooling and U-turn heat sink design) are evaluated and compared in terms of their End-of-Life (EOL) characteristics. The thesis concludes with a summary of the key findings, a discussion of limitations and an outlook for future research.

2 Fundamentals and state of the art

This chapter presents the theoretical background and state-of-the-art technologies relevant to the investigation of BTMS concepts. To provide the necessary context, an overview of BEV, VTMS and BTMS architectures is first introduced. Subsequently, the key principles of LIB technology are discussed, with a focus on electro-thermal modeling approaches. Finally, the chapter outlines the fundamentals of heat transfer in indirect liquid cooling systems.

2.1 Battery electric vehicle

The shift toward powertrain electrification in the automotive industry can be categorized into three main vehicle types, which are hybrid electric vehicles (HEVs), plug-in hybrid electric vehicles (PHEVs) and BEVs. While HEVs and PHEVs rely on a combination of internal combustion engines, typically powered by gasoline and electric machines supplied by batteries, BEVs rely exclusively on a traction battery as their power source to meet the vehicle's entire energy demand. [14]

In this thesis, particular emphasis is placed on the energy required at the LIB cell level of BEVs. This energy demand can be derived from a given velocity-time profile defined at the vehicle level and subsequently mapped down to the cell level. For this purpose, the total energy demand of the vehicle $P_{\mathrm{v,ed}}$ can be determined by

$$P_{\mathrm{v,ed}} = P_{\mathrm{acc}} + P_{\mathrm{air}} + P_{\mathrm{roll}} + P_{\mathrm{climb}} + P_{\mathrm{pt}} + P_{\mathrm{slip}} \tag{2.1}$$

where P_{acc} represents the power required for acceleration, P_{air} denotes power to overcome aerodynamic drag, P_{roll} corresponds to rolling resistance power and P_{climb} accounts for the power needed to overcome elevation changes. The terms P_{pt} and P_{slip} represent powertrain losses and slip losses, respectively. The total power consumption can be categorized into external losses ($P_{\mathrm{air}}, P_{\mathrm{roll}}, P_{\mathrm{acc}}, P_{\mathrm{climb}}$) and internal losses ($P_{\mathrm{pt}}, P_{\mathrm{slip}}$). [15]

M. Auch, *Holistic Modeling Framework for the Investigation of Innovative Battery Thermal Management Concepts*, Wissenschaftliche Reihe Fahrzeugtechnik Universität Stuttgart, https://doi.org/10.1007/978-3-658-52612-2_2

The energy demand of the traction battery $P_{\mathrm{tb,ed}}$ can be calculated from the total energy demand of the vehicle by

$$P_{\mathrm{tb,ed}} = \begin{cases} \frac{P_{\mathrm{v,ed}}}{\eta_{\mathrm{pt}}} & \text{for acceleration} \\ P_{\mathrm{v,ed}}\,\eta_{\mathrm{pt}} & \text{for regenerative braking} \end{cases} \tag{2.2}$$

dependent on the BEV operating condition. During acceleration, the battery must supply additional power to account for powertrain losses, resulting in a higher energy demand. Conversely, during regenerative braking, only a portion of the kinetic energy recovered from the drivetrain can be stored due to inefficiencies, resulting in reduced energy recovery. [16]

With the traction battery energy demand $P_{\mathrm{tb,ed}}$ determined, the battery pack architecture must be considered to calculate the resulting current at the individual cell level. The cell current I_{cell} can be approximated by

$$I_{\mathrm{cell}} = \frac{P_{\mathrm{tb,ed}}}{U_{\mathrm{N,cell}}\, n_{\mathrm{s}}\, n_{\mathrm{p}}} \tag{2.3}$$

where $U_{\mathrm{N,cell}}$ is the nominal voltage of a single cell, n_{s} is the number of cells connected in series and n_{p} is the number of parallel-connected strings within the battery pack. Using the nominal cell voltage represents a simplified assumption, as the actual voltage depends on factors such as the state of charge (SOC), temperature and applied current. Finally, the total energy demand of a BEV includes contributions from auxiliary systems, such as the VTMS and onboard electronics. However, these loads vary significantly depending on the specific system configuration and operating conditions of the vehicle [15].

2.2 Vehicle thermal management system

The VTMS is responsible for regulating the thermal operating ranges of the HVAC system, power electronics, electric machines and traction battery [17]. A central challenge in VTMS design lies in the fact that the optimal temperature ranges for these components differ significantly, as illustrated in Figure 2.1. Notably, both the vehicle cabin and the traction battery require much narrower temperature windows than the electric machines and power electronics, which must be carefully accounted for in system-level thermal strategies. To maximize overall vehicle efficiency, it is essential to minimize the energy consumption of the VTMS. This places special emphasis on the architecture of the BEV VTMS.

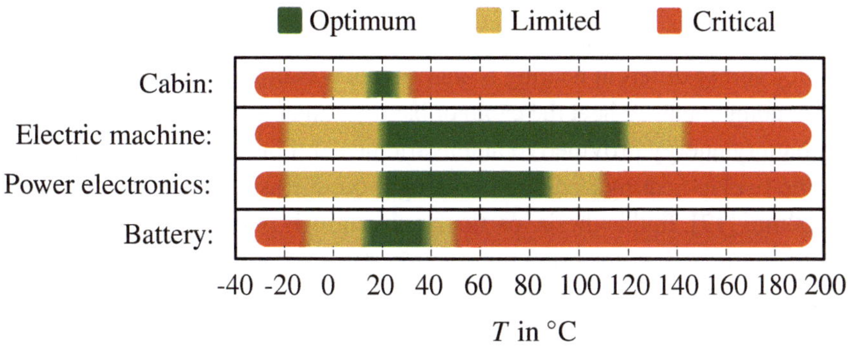

Figure 2.1: Schematic illustration of the optimal operating temperature ranges of key BEV components. Adapted from [18].

The passenger cabin temperature is regulated by a dedicated refrigeration circuit, whereas the battery, electric machines and power electronics are thermally managed using liquid coolant circuits. Early VTMS architectures in BEVs typically employed two separate coolant circuits, one dedicated to the traction battery and another to the electric machines and power electronics. However, with the increasing demand for extended driving range and improved energy efficiency, both vehicle-level and thermal management efficiency have become critical design objectives. Consequently, modern VTMS architectures employ integrated, energy-optimized configurations, such as Tesla's Octovalve system, which enable flexible thermal routing and improved heat reuse strategies [17].

The Octovalve system is named for its ability to control eight distinct fluid pathways through a single integrated valve assembly. Based on the thermal requirements of the cabin, battery, power unit and the ambient environmental conditions, the system dynamically selects the optimal valve configuration to maximize VTMS efficiency. This is achieved by coupling the coolant circuits of the battery, power electronics and electric machines under favorable operating conditions, rather than maintaining separate loops. However, this integration requires that all three components be thermally regulated by a shared coolant circuit, which introduces additional control complexity and necessitates careful management of their differing optimal temperature ranges [17].

The primary advantage of this system lies in its ability to improve efficiency across different heating scenarios, such as utilizing waste heat from other components to heat the cabin or battery, rather than relying exclusively on battery energy provided to an electric heater. For a detailed breakdown of the twelve heating and three cooling modes, it is referred to Wray and Ebrahimi [17]. In this thesis, particular emphasis is placed on the fast charging scenario and its corresponding VTMS design. For this purpose, only the heat removal path during fast charging is discussed for the Octovalve system illustrated in Figure 2.2.

The liquid coolant absorbs heat generated by the battery and transfers it to the refrigerant circuit via the chiller. Within the refrigerant loop, the waste heat is then rejected through two primary pathways. First, the heat is transferred from the refrigerant circuit to the coolant loop through a liquid-cooled condenser and subsequently dissipated to the ambient environment via the radiator. Second, when additional cabin heating is required, a portion of the recovered waste heat can be redirected to raise the cabin temperature, thereby enhancing overall system efficiency through thermal energy reuse. A key architectural characteristic relevant to this thesis is that both the inlet and outlet of the battery coolant loop are connected to a single integrated component, the Octovalve, which centrally manages thermal routing and flow direction.

In passenger BEVs, fast charging already presents one of the most demanding thermal management challenges. A critical limitation arises during vehicle standstill, where heat dissipation via the radiator and condenser is significantly reduced due to the restricted fan driven airflow. This issue is expected to become even more pronounced with the advancement toward extreme fast charging, which aims to further reduce charging times by substantially increasing the charging current [19]. Addressing this challenge

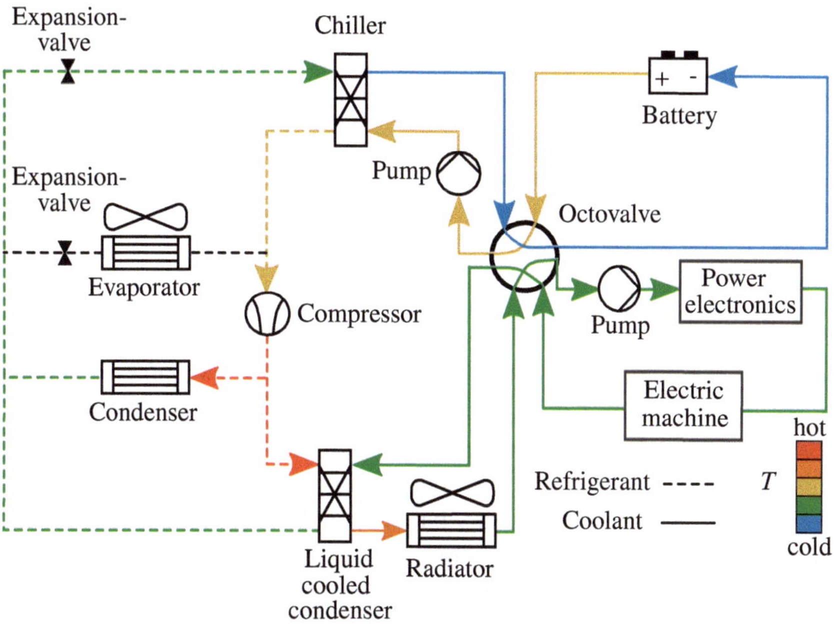

Figure 2.2: Tesla Octovalve-based VTMS architecture in cooling mode during fast charging. Adapted from [17].

requires effective cooling strategies on both the VTMS and BTMS levels. On the BTMS side, the system design must be capable of rapidly extracting heat from the battery to prevent excessive temperature rise and avoid the risk of thermal runaway. As charging power continues to increase, the performance of the BTMS becomes increasingly critical to ensuring battery longevity, safety and overall vehicle efficiency. To support this discussion, the following sections introduce the fundamentals of BTMS architecture and LIB cell behavior.

2.3 Battery thermal management system

This section provides an overview of the state of the art in BTMSs. To establish the necessary foundation, the working principles of cylindrical LIB cells are introduced, with particular emphasis on their electrical and thermal behavior. Subsequently, the heat transfer methods commonly employed in BTMS designs are discussed.

2.3.1 Fundamentals of lithium-ion batteries

BEVs utilize electrochemical energy storage systems as their primary energy source. These systems can be categorized into five main types, primary batteries (single-use non-rechargeable), secondary batteries (rechargeable), fuel cells (active material from external source), supercapacitors and photovoltaic devices. While primary/secondary batteries and fuel cells directly convert chemical energy into electrical energy, supercapacitors store energy through electrostatic and capacitive interactions. In contrast, photovoltaic devices generate electrical power by converting radiation energy. [20]

For BEV applications, secondary batteries (particularly LIBs), are the preferred choice due to their high energy density, high power density, low self-discharge rate and long service life [21].

Working principle

Generally, LIB cells operate by reversibly converting chemical energy into electrical energy during charging and discharging processes. A schematic representation of the internal layer structure during discharge is shown in Figure 2.3. The cell consists of six primary layers. These include the positive current collector, the positive electrode (cathode), the electrolyte, the separator, the negative electrode (anode) and the negative current collector. Additionally, during the initial charge-discharge cycles, two passivation layers form at the electrode-electrolyte interfaces. At the cathode-electrolyte interface, a cathode electrolyte interphase (CEI) layer develops, while at the anode-electrolyte interface, a SEI layer forms. These interphases, particularly the SEI, are critical for stabilizing the electrode-electrolyte interactions, suppressing further electrolyte decomposition and ensuring long-term cell performance and safety. [22, 23]

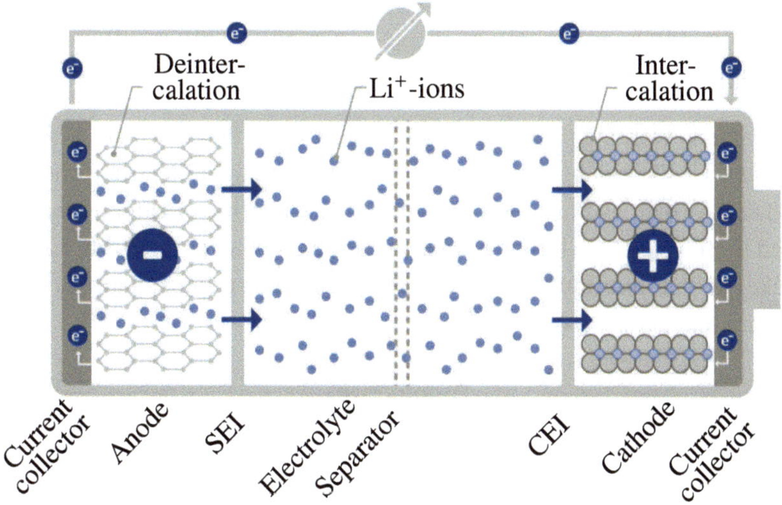

Figure 2.3: Schematic illustration of the structure and working principle of a LIB cell during discharge. Adapted from [22].

The working principle of LIB cells is based on the reversible intercalation and deintercalation of lithium ions within the electrodes (active material), driven by electrochemical redox reactions. The positive electrode and negative electrode are connected via an ionically conductive electrolyte, which enables the migration of lithium ions between the electrodes. To prevent direct contact between the electrodes and thus avoid an internal short circuit, a separator is placed between them. Each electrode is connected to a current collector, which facilitates electron flow. During charging and discharging, lithium ions deintercalate from one electrode and migrate through the electrolyte to the other electrode, while the associated electron flow occurs externally through the load or charging device. [22]

LIB cells provide a very good charge conservation, indicated by the often taken assumption of a coulombic efficiency of $\eta_c = 1$, which is the ratio between discharge and charge capacity within the same cycle [24]. For the investigation of LIB cells, it is important to define the current sign convention. In this thesis, discharging is defined as a positive current, while charging is defined as a negative current.

Formats and chemistry

The general working principle of LIB cells remains the same across all types. However, LIBs can be further classified based on their electrode chemistry and cell format, as illustrated in Figure 2.4. The primary cathode materials used in BEVs include lithium nickel manganese cobalt oxide (NMC), lithium nickel cobalt aluminum oxide (NCA), lithium iron phosphate (LFP) and lithium manganese oxide (LMO).

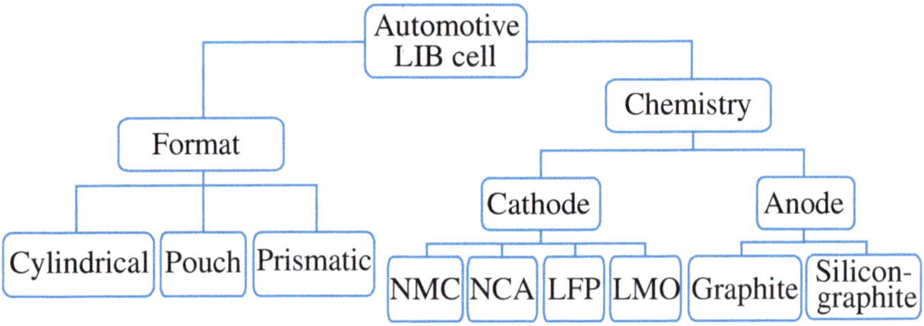

Figure 2.4: Cell format and cathode/anode chemistry classifications used in automotive LIB cells.

The industry trend for cathode chemistry is shifting towards two dominant options, NMC and LFP [25]. On one hand, high-nickel NMC offers high energy density and is preferred for high-performance BEVs. On the other hand, cobalt-free LFP is valued for its low cost, safety and suitability for budget-oriented BEVs. For the anode, graphite has long been the standard material. However, the industry is increasingly adopting silicon-graphite composites, which enhance energy density due to silicon's higher lithium storage capacity. Nevertheless, higher silicon content can lead to severe volume variation, which reduces the LIB cells life-cycle [26].

Furthermore, LIB cells can be classified according to their formats, which are cylindrical, prismatic, pouch and coin cells. Each format consists of multiple stacked layers of the fundamental layered structure. In these formats, the positive and negative current collector layers are connected to one positive and one negative terminal. This simplifies the connection of multiple LIB cells to achieve the power demands required in BEVs. While coin cells are not commonly used in BEVs, cylindrical, prismatic and pouch cells are all found in commercial BEVs. In pouch cells, the layers are stacked. In cylindrical cells, they are rolled into a spiral, forming the so-called jelly roll structure. Prismatic cells can utilize either stacking or rolling configurations.

Due to their mass-production scaling ability, cylindrical LIB cells are becoming increasingly popular in BEV applications [27]. The latest example being BMW shifting from a prismatic to a cylindrical cell format for their 'Neue Klasse' vehicles [28]. Consequently, this thesis focuses on the cylindrical cell format. Understanding the operating behavior of LIB cells is essential for BEV system design. In particular, both electrical and thermal performance must be considered. The following sections provide an overview of the fundamentals and state-of-the-art modeling approaches for the electrical and thermal behavior of cylindrical LIB cells.

2.3.2 Electrical modeling of lithium-ion batteries

The purpose of electrical models for LIB cells is to simulate their electrical behavior over all operating conditions. To achieve this, it is essential to first introduce the most important macroscopic properties of LIB cells.

The C-rate is a parameter used to describe the charge/discharge current rate applied to a LIB cell and is defined as

$$\text{C-rate} = \frac{|I|}{Q_\mathrm{N}} \tag{2.4}$$

allowing for a standardized comparison of relative current loads across cells of different capacities. The SOC represents the ratio of the remaining actual charge Q_act to the nominal capacity Q_N of the LIB cell and is given by

$$\mathrm{SOC} = \frac{Q_\mathrm{act}}{Q_\mathrm{N}} 100 \tag{2.5}$$

expressed as a percentage. Another key parameter is the Open Circuit Voltage (OCV), which represents the equilibrium potential of the cell when no external current is applied. The OCV arises from the difference in equilibrium potential of cathode and anode and changes with SOC. The relationship between OCV and SOC is a fundamental characteristic of a LIB cell. Figure 2.5 illustrates the OCV curves for charging and discharging. A noticeable effect is that the OCV-SOC curves do not perfectly overlap, a phenomenon referred to as OCV hysteresis. This effect is caused by electrochemical and kinetic processes within the electrodes and contributes to reduced cyclic energy efficiency. Additionally, OCV varies with chemistry and temperature, further influencing LIB performance. [24]

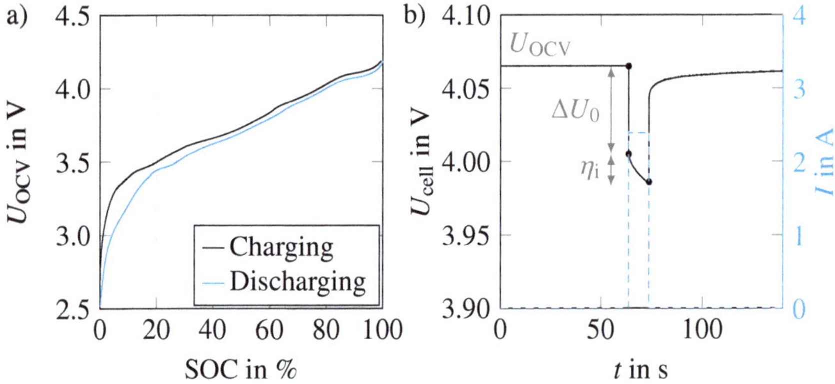

Figure 2.5: a) OCV during charge and discharge over SOC; b) Terminal voltage response to an applied discharge pulse. Adapted from [29].

As illustrated in Figure 2.5 b), when a load is applied to a LIB cell, the terminal voltage U_{cell} deviates from the OCV voltage U_{OCV}. This deviation can be calculated by

$$U_{cell} = U_{OCV} - \Delta U_0 - \sum_i \eta_i \tag{2.6}$$

and arises due to the internal ohmic resistance voltage drop ΔU_0 and polarization effects η_i. During discharge, the LIB cell must compensate for these losses, resulting in a voltage drop at the terminals. During charging, the applied voltage must overcome the internal resistance and polarization effects to drive the electrochemical reactions. This voltage behavior is significantly influenced by several factors. These include the electrode chemistry, operating temperature, current rate (C-rate), SOC and state of health (SOH). Consequently, an appropriate electrical model is required to accurately capture and predict the dynamic behavior of the terminal voltage U_{cell} under varying load conditions. [24, 30]

Electrical model

There are three commonly used modeling approaches for the electrical behavior of LIB cells, as shown in Figure 2.6. These can be classified as electrochemical or physics-based models (Single-Particle Model, Doyle-Fuller-Newman, Pseudo-2D Model), empirical or data-driven models and equivalent circuit models (Rint Model, Thevenin Model, Dual Polarization Model). Physics-based models offer the highest accuracy across a wide range of operating conditions. They employ physical equations to describe the electrochemical transport operations occurring within the cell and typically require a large number of parameters for accurate implementation. However, the complexity of the governing equations also results in high computational cost. [31, 32]

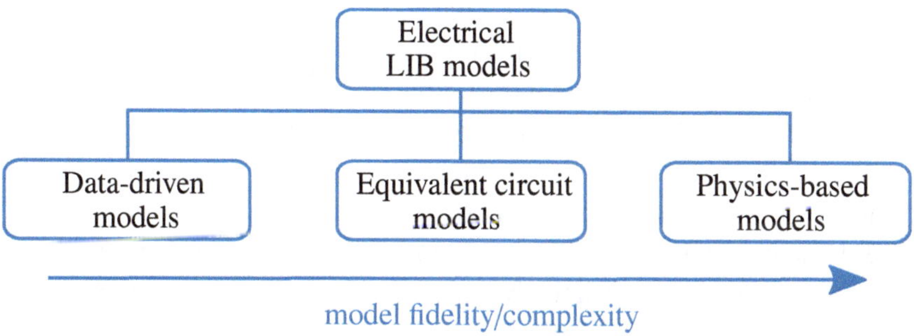

Figure 2.6: Classification of electrical modeling approaches for LIB cells.

In contrast, data-driven and equivalent circuit model (ECM)-based modeling approaches do not rely on the underlying electrochemical equations. Instead, they are parameterized using experimental data. Data-driven models typically employ machine learning techniques and require extensive testing to adequately capture the full operating range of the battery. However, their predictive accuracy is inherently limited by the quality of the training data [33]. ECMs, on the other hand, represent the battery using simplified analogies to electrical circuits. Their parameterization is supported by well-established test protocols, offering a favorable trade-off between experimental effort, computational cost and sufficient modeling accuracy. Moreover, due to their good scalability from cell to pack level and straightforward coupling with thermal models, ECMs remain the predominant choice for electro-thermal investigations in BTMS applications [34].

ECMs can be classified into two categories, which are time- and frequency-domain models [35]. The primary distinction between them lies in the experimental procedures used for model parameterization. Frequency-domain ECMs rely on a more complex experimental technique known as electrical impedance spectroscopy (EIS). This method excites the LIB cell with signals over a wide frequency range and records its voltage and current responses. The data obtained through EIS not only supports parameterization across a broad range of operating conditions but also provides deeper insight into the cell's electrochemical processes and degradation behavior. However, in addition to the more advanced test setup and parameter estimation complexity, EIS typically requires significantly longer testing durations compared to time-domain methods. In contrast, time-domain ECMs utilize standard battery laboratory equipment and are parameterized using step or pulse current profiles while recording current, voltage and temperature. While frequency-domain ECMs are primarily used for detailed cell diagnostics, time-domain ECMs are widely applied in BTMS research and system-level simulations due to their relatively simple parameterization process, lower cost and shorter test times. [36]

In Figure 2.7, the state of the art in time-domain ECM modeling is illustrated for an ECM with two resistance-capacitance (RC)-networks (Dual Polarization Model) using the analogy to electrical circuits. The model consists of the OCV U_{OCV}, an instantaneous ohmic resistance voltage drop ΔU_0 and two dynamic polarization voltage drops ΔU_{RC_n}, each represented by an RC-network. With

$$U_{\mathrm{cell}} = U_{\mathrm{OCV}} - \Delta U_0 - \sum_{i=1}^{2} \Delta U_{\mathrm{RC}_i} \tag{2.7}$$

the terminal cell voltage U_{cell} can be calculated.

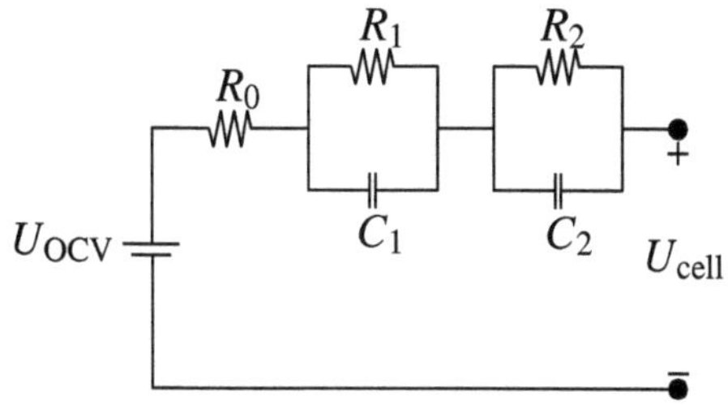

Figure 2.7: Schematic representation of a Dual Polarization ECM featuring two RC-networks for modeling the electrical behavior of LIB cells.

For this purpose, the ohmic resistance voltage drop is defined as $\Delta U_0 = I R_0$ and by using Kirchhoff's law

$$\begin{bmatrix} U_{\mathrm{RC},1} \\ U_{\mathrm{RC},2} \\ \mathrm{SOC} \end{bmatrix}_i = \begin{bmatrix} \exp\left(-\frac{\Delta t}{R_1 C_1}\right) & 0 & 0 \\ 0 & \exp\left(-\frac{\Delta t}{R_2 C_2}\right) & 0 \\ 0 & 0 & 1 \end{bmatrix} \begin{bmatrix} U_{\mathrm{RC},1} \\ U_{\mathrm{RC},2} \\ \mathrm{SOC} \end{bmatrix}_{i-1} + \begin{bmatrix} R_1 \left(1 - \exp\left(-\frac{\Delta t}{R_1 C_1}\right)\right) \\ R_2 \left(1 - \exp\left(-\frac{\Delta t}{R_2 C_2}\right)\right) \\ \frac{\Delta t}{Q_{\mathrm{N}}} \end{bmatrix} I_{\mathrm{cell}} \tag{2.8}$$

the polarization voltage drops can be determined dependent on the solution of the previous time step t_{i-1} [37].

The accuracy of ECMs can be improved by using more RC-networks in series. However, according to Hu et al. [38], the increased accuracy of more than two RC-networks does not overcome the increased computational effort. Model parameters can typically be identified using simple pulse current tests under varying conditions, including temperature, SOC, hysteresis and current rates [29, 30]. Moreover, these parameters change over the life cycle of the LIB cell as a result of degradation mechanisms associated with aging [30]. Consequently, the following section introduces the fundamental processes of LIB aging.

Aging of LIBs

In general, LIB aging refers to the macroscopic effects of capacity fade and internal resistance growth. It can be categorized into two types, which are calendar and cycle aging. Calendar aging occurs even when LIB cells are not under load, primarily influenced by SOC and temperature. This is largely due to the formation and growth of the SEI-layer on the anode [39, 40]. On the other hand, cycle aging occurs during the operation of the LIB cells and is influenced by the current rate (C-rate), SOC and operating temperature. The aging behavior of LIB cells can be categorized according to the SOH-charge throughput behavior, as illustrated in Figure 2.8 a). There are three common aging patterns, which are linear, superlinear and sublinear [41, 42]. For LIB cells in BEVs, maintaining sublinear aging behavior is essential to extend lifespan.

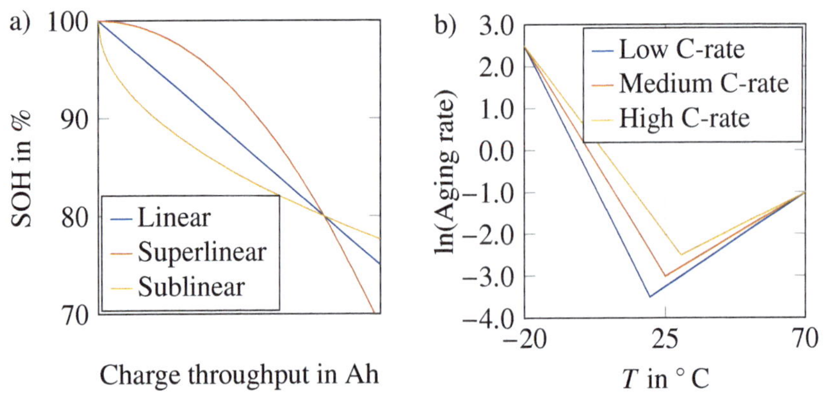

Figure 2.8: a) Aging behavior of LIB cells (Adapted from [41]); b) Influence of C-rate on the aging rate of LIB cells (Adapted from [42]).

Since aging significantly influences the LIB cells performance over their life-cycle, in the application of BEVs two common criteria are used to define the EOL. These two criteria are either when its capacity degrades to SOH = 80 % or when its internal resistance doubles (200 % of its initial Begin-of-Life (BOL) value) [43]. There are three driving forces for battery aging, which are the loss of lithium inventory, the loss of active anode material and the loss of active cathode material [44]. While LIB aging involves multiple degradation mechanisms, this thesis specifically examines temperature-related aging effects, as they play a dominant role in long-term performance. As a result, it is referred to [41, 44] for a detailed description of LIB degradation effects.

The operating temperature is a crucial parameter to maximize the life and performance of LIB cells. According to Waldmann et al. [6], the optimum operating temperature for LIB cells is around $T \approx 25\,°\mathrm{C}$. Generally, operating the LIB cell at temperatures above the optimum temperature leads to accelerated cathode degradation and SEI-layer growth on the anode becomes the dominating aging mechanism [6, 7]. Conversely, operating at temperatures below the optimum, the dominant aging mechanism is lithium plating [6]. Additionally, as illustrated in Figure 2.8 b), higher charging C-rates accelerate LIB cell aging due to increased SEI-layer growth and lithium plating [8, 42]. The C-rate can further influence the optimal operating temperature, as higher C-rates demand faster reaction kinetics, which are more efficient at elevated temperatures [8, 42]. Finally, while the average cell temperature plays a dominant role in predicting aging behavior, temperature gradients within a cell can contribute to accelerated degradation effects [45].

Heat generation

As with any energy conversion process, both the transformation from electrical to chemical energy during charging and from chemical to electrical energy during discharging are accompanied by energy losses. These losses result primarily in waste heat. The total heat generation within a LIB cell can be categorized based on the underlying physical mechanisms and their sources. According to Bernardi et al. [46], the total heat generation rate can be expressed as

$$\dot{Q}_{\text{tot}} = \dot{Q}_{\text{irr}} + \dot{Q}_{\text{rev}} + \dot{Q}_{\text{reac}} + \dot{Q}_{mix} \tag{2.9}$$

consisting of the irreversible ohmic heating losses $\dot{Q}_{\text{irr}}$, the reversible heat associated with the entropy change of the electrochemical reactions $\dot{Q}_{\text{rev}}$, the enthalpy changes due to mixing effects in the electrolyte $\dot{Q}_{\text{mix}}$ and the heat generation due to the side reactions $\dot{Q}_{\text{react}}$.

The heat contributions from side reactions $\dot{Q}_{\text{react}}$ ([47]) and mixing effects in the electrolyte $\dot{Q}_{\text{mix}}$ ([48]) are generally negligible under standard operating conditions of LIB cells. As a result, the total heat generation equation can be simplified to

$$\dot{Q}_{\text{tot}} = \underbrace{I\,(U_{\text{OCV}} - U_{\text{cell}})}_{\dot{Q}_{\text{irr}}} - \underbrace{I\,T\,\frac{\partial\,U_{\text{OCV}}}{\partial\,T}}_{\dot{Q}_{\text{rev}}} \tag{2.10}$$

where U_{cell} denotes the terminal voltage of the cell and ${}^{\partial U_{\text{OCV}}}/_{\partial T}$ is the LIB entropic coefficient [46]. While the irreversible heat generation is always positive, the reversible heat generation can be either positive or negative depending on the SOC [29]. This expression is widely used in the literature for estimating heat generation in LIB cells, due to its analytical simplicity and practical relevance [34]. However, it is important to highlight a key limitation. Since the current I appears as a direct multiplier in both terms, the calculated heat generation becomes zero as soon as the current drops to zero. As a result, residual heat generation due to prior polarization effects is neglected [49].

Both the internal resistance and the reversible heat generation of a LIB cell are temperature-dependent, requiring an electro-thermal coupled approach. In such models, the electrical model calculates the generated heat based on the operating conditions at each time step, while the thermal model determines the resulting cell temperature. The two models are updated iteratively at each time step, allowing for dynamic interaction between electrical and thermal behavior. As a result, reliable LIB modeling demands accurate prediction of both the electrical and thermal responses of the cell under varying operating conditions. To support this, the following section presents the state of the art in thermal modeling approaches for LIB cells.

2.3.3 Thermal modeling of lithium-ion batteries

The general internal structure of a cylindrical LIB cell is illustrated in Figure 2.9. Developing accurate thermal models requires not only a detailed understanding of the jelly roll layers but also of the overall cell design. This is particularly important when heat transfer methods or BTMS concepts are applied, as additional LIB cell structural components can significantly influence the heat flow path and thermal behavior. In addition to the jelly roll, key structural elements such as the cell casing, cap and current collectors (positive and negative tabs) play a crucial role in heat generation and dissipation. Most commercially available high-energy cylindrical cells feature a single positive and negative tab for electrical connection. However, recent developments have introduced tab-less cylindrical designs, in which the single tab is replaced by multiple tabs distributed along the jelly roll edges, connecting directly to the positive and negative terminals [50]. This design innovation reduces thermal and electrical resistance, thereby improving both heat dissipation and current distribution [50].

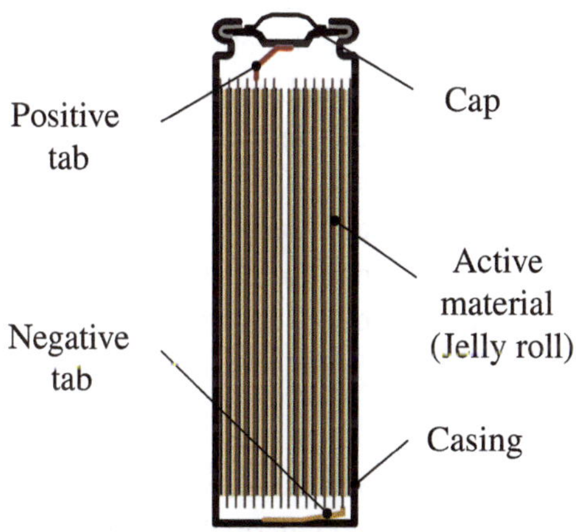

Figure 2.9: Cross-sectional view of a cylindrical LIB cell, illustrating key internal components.

Cylindrical LIB cells are characterized by an axisymmetric structure (apart from the tab locations), which makes it advantageous to employ a cylindrical coordinate system (r, ϕ, z) for thermal analysis. This coordinate system simplifies the mathematical modeling of heat transfer and enables analytical or semi-analytical solutions for thermal properties such as density, specific heat capacity and thermal conductivity. The effective thermal properties of a cylindrical LIB cell can be estimated by considering the volume-weighted contributions of its components [51]. Notably, the thermal conductivity in the radial direction follows a series configuration of the jelly roll layers, resulting in lower effective conductivity and slower heat transfer. In contrast, the axial and circumferential directions approximate a parallel configuration, allowing for more efficient heat dissipation. Consequently, cylindrical LIB cells exhibit highly anisotropic thermal behavior, with directional thermal conductivity differences often exceeding an order of magnitude [51].

Using the effective thermal properties, the energy conservation equation for anisotropic heat conduction in cylindrical coordinates is expressed as

$$c_{\mathrm{p}} \rho \frac{\partial T}{\partial t} = \frac{1}{r} \frac{\partial}{\partial r} \left(r k_{\mathrm{r}} \frac{\partial T}{\partial r} \right) + \frac{1}{r^2} \frac{\partial}{\partial \phi} \left(k_{\phi} \frac{\partial T}{\partial \phi} \right) + \frac{\partial}{\partial z} \left(k_{\mathrm{z}} \frac{\partial T}{\partial z} \right) + \dot{q} \tag{2.11}$$

where c_{p} is the effective specific heat capacity, ρ is the effective density, T is the temperature, k denotes the directional thermal conductivity (with subscripts r, ϕ, z referring to the radial, circumferential and axial directions) and $\dot{q}$ is the volumetric heat generation rate.

Depending on the desired level of detail and computational resources, the optimal thermal modeling approach for LIB analysis may vary [34]. In the following, the most commonly used three-dimensional (3D) and one-dimensional (1D) thermal modeling strategies are introduced.

3D thermal model

The most accurate prediction of the internal temperature distribution in LIB cells can be achieved using 3D thermal models. These models require detailed knowledge of the internal structure of the LIB cell as well as the thermal properties of each individual component. Existing 3D modeling approaches can be categorized based on two criteria, which are the level of resolution of the jelly roll and the resolution of the remaining structural components. This classification is illustrated in Figure 2.10.

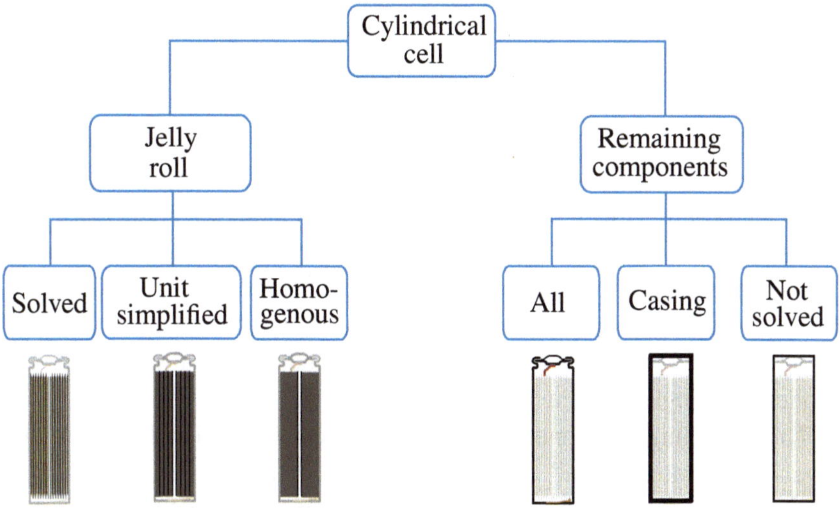

Figure 2.10: Categorization and cross-sectional views of different levels of geometric detail in 3D thermal LIB models.

There are three common modeling approaches for representing the jelly roll structure in cylindrical LIB cells. These include explicitly resolving the individual jelly roll layers, simplifying the structure into repetitive units of the layers and assuming a homogeneous cylindrical core. The most detailed method, explicitly resolving the jelly roll layers, is primarily beneficial in high-fidelity three-dimensional electrochemical-thermal models. This approach enables the temperature-dependent losses and corresponding heat generation to be applied to each individual component layer of the jelly roll. However, it requires a comprehensive data-set of electrochemical and thermal properties for the specific LIB cell and results in significantly higher computational cost. [52]

The unit-based simplification approach allows for the consideration of temperature gradients within the jelly roll and their impact on the electrochemical behavior of the LIB cell. In this method, heat generation is calculated individually for each jelly roll unit, taking into account local temperature variations. Although the internal structure is simplified, the model still captures important thermal-electrochemical coupling effects by using reduced-order electrochemical or electrical models to determine local heat generation. Commonly used simplified models in this context include the Pseudo-2D model, the Single Particle Model and ECMs. [52, 53]

Finally, the homogeneous approach assumes uniform heat generation throughout the entire jelly roll, based solely on the average temperature of the cell. In this method, three-dimensional temperature gradients within the jelly roll arise exclusively from the application of effective thermal properties, rather than spatially resolved heat sources. Similar to the unit-based approach, the temperature-dependent heat generation can be calculated using simplified electrical or electrochemical models such as the Pseudo-2D model, the Single Particle Model or ECMs [52].

The remaining components of the LIB cell can be modeled with varying levels of geometric detail. These approaches can be categorized into fully resolving all components [54, 55], modeling only the LIB casing [56] or neglecting the remaining components entirely [57, 58, 59]. In models that resolve all components, the cell casing, cap structure and both current collector tabs are explicitly included in the geometry. It is important to note that the jelly roll is the only region within the LIB cell where both irreversible and reversible heat generation occurs. In contrast, the remaining structural components contribute solely through irreversible ohmic losses.

Consequently, in addition to the decomposition of total heat generation into electrochemical contributions, the heat generation within a LIB cell can also be categorized based on its source between the remaining components within the cell structure. As demonstrated by Robinson et al. [60], spatially assigning ohmic resistances, and thus heat sources, to the corresponding structural components is crucial for accurately capturing the internal temperature distribution of a cylindrical LIB cell. Following this approach, the total heat generation rate of the battery cell $\dot{Q}_{\mathrm{tot}}$ can be partitioned as

$$\dot{Q}_{\mathrm{tot}} = \dot{Q}_{\mathrm{jr}} + \dot{Q}_{\mathrm{cap}} + \dot{Q}_{\mathrm{pt}} + \dot{Q}_{\mathrm{nt}} \tag{2.12}$$

where $\dot{Q}_{jr}$, $\dot{Q}_{cap}$, $\dot{Q}_{pt}$, $\dot{Q}_{nt}$ represent the heat generated in the jelly roll, the cell cap, the positive terminal and the negative terminal, respectively. While this component-wise discretization increases modeling complexity and computational cost, it significantly improves the accuracy of temperature field predictions across the entire cell. This is particularly important in the context of thermal management, as the heat flow from the cooling channels to the jelly roll can directly be affected by the thermal and geometric characteristics of the surrounding cell components.

A common simplification is to consider only the LIB cell casing, while neglecting internal structural components such as the cell cap and terminals. This is often justified by the fact that most thermal management strategies are applied to the base or cell mantle of the LIB cells, making the influence of the cap design secondary. In many three-dimensional battery module simulations focused on cooling design optimization, the remaining components are not modeled at all. Instead, the entire LIB cell is often represented as a simplified cylindrical volume with homogenized thermal properties. While this approach significantly reduces computational cost and is suitable for preliminary geometric optimization, it can lead to misleading conclusions when evaluating cooling performance. In particular, when analyzing the effect of cooling channel position or direction, such simplifications may fail to capture the true thermal behavior and average cell temperature.

While 3D thermal models offer superior accuracy in capturing the spatial temperature distribution within LIB cells, their application becomes impractical when investigating life-cycle characteristics and aging effects on the full module or pack level due to the associated high computational cost. As a result, lumped thermal models (1D thermal models) are often the preferred choice for such studies, providing a favorable trade-off between computational efficiency and acceptable modeling accuracy.

1D lumped thermal model

Similar to 3D models, 1D lumped models can be constructed with varying levels of detail, depending on the application requirements. In general, 1D thermal models of LIB cells are based on the thermal mass principle, wherein the cell is represented as a point mass with lumped thermal properties. Although the underlying formulation corresponds to a zero-dimensional (0D) physical model, these models are referred to as 1D in this thesis due to the spatial resolution achieved by connecting multiple thermal nodes along the modeled system.

As illustrated in Figure 2.11, two primary approaches are commonly used for 1D thermal modeling of LIB cells. These are the single-node thermal model and the thermal resistance network model [37]. The single-node thermal model represents the entire LIB cell as a single thermal node, requiring only the assignment of effective thermal properties such as specific heat capacity, mass and density. A key advantage of this approach lies in its excellent scalability, allowing entire cells, modules or even complete battery packs to be aggregated into a single thermal node. Due to its low computational cost, it is particularly well-suited for large-scale or real-time simulations, such as system-level studies involving the VTMS.

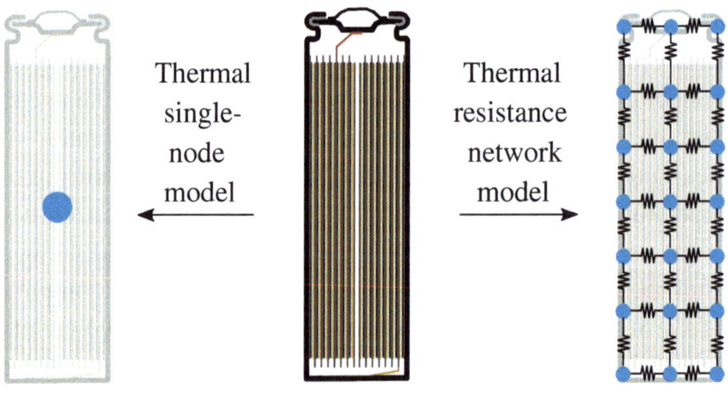

Figure 2.11: Categorization of different levels of 1D thermal LIB models.

A higher level of detail in 1D thermal modeling can be achieved using the thermal resistance network model [54]. In this approach, the LIB cell is discretized into multiple thermal mass points, whose number and placement can be adjusted to reflect the internal structure of the cell with greater fidelity. These thermal masses are interconnected by thermal resistances that represent both heat conduction paths and thermal interfaces between adjacent regions. This modeling technique enables the evaluation of spatial temperature distributions within the cell, with the resolution governed by the number and location of the thermal nodes. Moreover, analogous to the previously described 3D thermal models, localized heat generation can be incorporated by assigning heat sources to specific nodes representing distinct LIB cell components [54].

Compared to the single-node thermal model, the thermal resistance network approach requires more detailed input data, including the individual component masses, material properties and heat transfer areas. However, it provides a more accurate representation of the internal thermal behavior, making it particularly well-suited for analyzing thermal gradients and heat propagation, especially in the context of applied heat transfer designs on the BTMS level.

2.3.4 Heat sink design

As previously discussed, the operating temperature plays a critical role in the electro-thermal-aging behavior of LIB cells. To manage thermal conditions and prevent excessive heating or cooling, a dedicated BTMS is employed. The BTMS is responsible for maintaining LIB cells within their optimal operating temperature range under varying load conditions and environmental influences. In addition, the BTMS must also minimize temperature variations both within individual cells and across the entire battery module or pack. Even small temperature gradients can lead to uneven aging, performance degradation and reduced overall battery lifespan [61]. As noted by Pesaran [5], a widely accepted design guideline is to limit temperature gradients, both within a single cell and between cells in a module or pack, to less than 5 K to mitigate these effects.

As illustrated in Figure 2.12, BTMS architectures can be classified according to their heat transfer method into passive, active and hybrid cooling approaches [9]. Passive systems include air cooling, phase change materials (PCM) and heat pipes (HP). Active systems also encompass air cooling, but primarily rely on liquid or refrigerant-based cooling methods. Hybrid cooling refers to any combination of these approaches. Early generations of BEVs predominantly used air cooling due to its simplicity and low cost [11]. However, as BTMS requirements have increased, particularly under fast charging conditions, air cooling has proven inadequate. Consequently, the industry has largely adopted liquid cooling, which provides more effective heat dissipation and improved temperature control [10, 11].

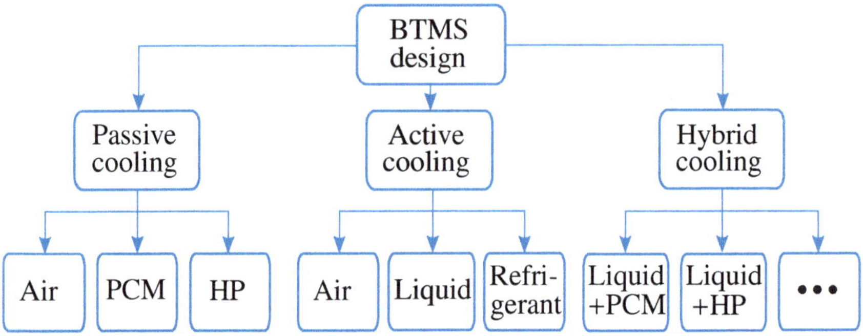

Figure 2.12: Classification of heat transfer methods across different BTMS concepts.

Liquid cooling strategies can be categorized into direct (immersion) and indirect liquid cooling. In direct cooling approaches, the coolant comes into direct contact with the LIB cells, necessitating the use of dielectric fluids to prevent electrical short circuits. In contrast, indirect liquid cooling relies on cooling plates or channels with a circulating coolant, transferring heat from the battery without direct fluid-cell contact. These plates are typically made from thermally and electrically conductive materials, requiring an additional layer of electrical insulation, either around the cooling channel or the LIB cells, to avoid unintended integration into the electrical system. While immersion cooling is a subject of ongoing research due to its high thermal efficiency, indirect liquid cooling has become the prevailing industry standard. This is primarily due to its practical balance between thermal performance, safety, system complexity and cost. [62, 63, 64]

In indirect liquid cooling designs, large temperature differences between the coolant and the LIB cells can lead to increased thermal gradients within individual cells, causing accelerated degradation over time [45]. Consequently, the coolant temperature should remain close to the operating temperature of the LIB cells, underscoring the importance of effective heat dissipation.

Furthermore, the choice of the coolant itself plays a critical role in ensuring reliable BTMS operation. Given that BEVs must remain functional in both hot and cold climates, the coolant must exhibit a low freezing point to prevent phase change and system failure in cold temperatures. While water offers excellent thermal conductivity and specific heat capacity, its relatively high freezing point of 0 °C limits its applicability in colder environments. To address this limitation while preserving favorable heat transfer characteristics, most BTMS designs commonly use a 50:50 mixture of ethylene glycol (EGL) and water. This effectively lowers the freezing point and is widely adopted in commercial BEVs. [64]

Finally, BTMS design requires an optimal trade-off between thermal performance (heat transfer, temperature control and uniformity) and cost factors such as design complexity and pressure drop.

2.4 Fundamentals of heat transfer

To investigate BTMS concepts, the relevant thermodynamic principles, particularly those governing heat transfer and pressure loss, are briefly summarized in the following. Fundamentally, three primary modes of heat transfer exist, which are heat conduction, heat convection and heat radiation.

Heat conduction refers to the transfer of thermal energy through a material without macroscopic fluid motion. It is governed by Fourier's law, which expresses the local heat flux density as

$$\dot{q}_{\mathrm{cond}} = -k \frac{\mathrm{d}T}{\mathrm{d}x} \tag{2.13}$$

where k is the thermal conductivity of the material (which may vary with temperature and pressure) and ${}^{\mathrm{d}T}/_{\mathrm{d}x}$ the temperature gradient over the spatial distance dx. The corresponding thermal resistance due to conduction through a wall of uniform thickness is defined as

$$R_{\text{th}} = \frac{th}{A\,k} \tag{2.14}$$

where th is the thickness of the material and A is the heat transfer cross-sectional area. [65]

Heat convection involves the transfer of heat between a solid and a fluid. This phenomena can be divided into natural and forced convection. Hereby, natural convection results from the movement of the fluid driven by density changes because of present temperature differences. Forced heat convection is driven by the movement of the fluid, resulting from an external force. The local heat flux density is described by Newton's law of cooling as

$$\dot{q}_{\text{conv}} = h_{\text{conv}}\,\Delta T \tag{2.15}$$

where h_{conv} is the convective heat transfer coefficient and ΔT is the temperature difference between the wall surface and the bulk fluid. [65]

Heat radiation occurs due to thermal energy carrying electromagnetic waves. The heat flux density due to heat radiation is calculated as

$$\dot{q}_{\text{rad}} = \epsilon\,\sigma\,(T_2^4 - T_1^4) \tag{2.16}$$

where ϵ represents the emissivity of the material, σ is the Stefan-Boltzmann-constant and T_1 and T_2 are the temperatures of the ambient and surface, respectively [66]. By using

$$h_{\text{rad}} = \epsilon\,\sigma\,\frac{(T_2^4 - T_1^4)}{(T_2 - T_1)} \tag{2.17}$$

a heat transfer coefficient can be approximated assuming heat radiation between two surfaces having the same emissivity [67].

Besides the fundamental heat transfer phenomena, this thesis places special emphasis on the flow behavior inside cooling channels. For this purpose, the dimensionless Nusselt number is introduced as a common characterization metric for the heat transfer. The Nusselt number quantifies the ratio of convective to conductive heat transfer and is defined as

$$Nu = \frac{d_{\text{h}}}{k_{\text{f}}}\,h_{\text{conv}} \tag{2.18}$$

where d_{h} is the hydraulic diameter of the cooling channel cross section, k_{f} represents the fluid thermal conductivity at the bulk-mean temperature and h_{conv} is the convective local or average heat transfer coefficient. To investigate the local heat transfer characteristics, the definition of the local convective heat transfer coefficient is defined as

$$h_{\mathrm{conv}}(z) = \frac{1}{A(z)} \frac{q(z)}{\sum_{x,y}[T_{\mathrm{w}}(x,y,z) - T_{\mathrm{m}}(z)]\,\mathrm{d}A(x,y,z)} \tag{2.19}$$

where the local heat transfer area is given as

$$A(z) = \sum_{x,y} \mathrm{d}A(x,y,z) \tag{2.20}$$

and the local heat flux

$$q(z) = \sum_{x,y} q(x,y,z)\mathrm{d}A(x,y,z) \tag{2.21}$$

over the wall surface. Further, $T_{\mathrm{w}}(x,y,z)$ represents the local wall temperature and $T_{\mathrm{m}}(z)$ is defined as the local fluid bulk-mean temperature which can be calculated by

$$T_{\mathrm{m}} = \frac{\int_{A_{\mathrm{c}}} \rho\, v\, T\,\mathrm{d}A_{\mathrm{c}}}{\int \rho\, v\,\mathrm{d}A_{\mathrm{c}}} \tag{2.22}$$

at the channel length position z. Finally, the average Nusselt number Nu_{avg} defined as

$$Nu_{\mathrm{avg}}(z) = \frac{d_{\mathrm{h}}}{k_{\mathrm{f}}} \frac{1}{\sum_{x,y,z} \mathrm{d}A(x,y,z)} \cdot \frac{\sum_{x,y,z} q(x,y,z)\mathrm{d}A(x,y,z)}{\sum_{x,y,z}[T_{\mathrm{w}}(x,y,z) - T_{\mathrm{m}}(z)]\mathrm{d}A(x,y,z)} \tag{2.23}$$

can similarly be computed. [68]

Additionally, the Nusselt number correlation for natural convection around a cylinder is defined as

$$Nu = \left\{0.752 + 0.387\,\left[Ra\, f_3(Pr)\right]^{1/6}\right\}^2 \tag{2.24}$$

dependent on the Rayleigh number Ra, Prandtl number Pr and the function f_3 [69]. Finally, the correlation for natural convection around a vertical plane plate is given as

$$Nu = \left\{0.825 + 0.387\ [Ra\, f_1(Pr)]^{1/6}\right\}^2 \tag{2.25}$$

dependent on the function f_1, as specified in Klan [69].

The occurring friction-related pressure losses in a pipe can generally be calculated by the Darcy-Weisbach equation

$$\Delta p = f \frac{l}{d_\mathrm{h}} \frac{\rho v_\mathrm{m}^2}{2} \tag{2.26}$$

where f represents the flow regime dependent Darcy friction factor, l the channel length and v_m the mean fluid velocity. [70]

3 Research objective

Building on the state of the art and fundamental concepts in BTMS, this chapter defines the research objective of this thesis. To put the objective into context, the limitations of existing BTMS approaches are first discussed. Subsequently, the two research questions are formulated and put into context within the scope of this thesis.

3.1 Limitations of the state of the art

As previously highlighted, the state of the art as heat transfer method in BTMS concepts is indirect liquid cooling. Despite its favorable trade-off between heat removal and required power demand, indirect liquid cooling presents a fundamental challenge that affects thermal homogeneity and battery pack performance, as illustrated in Figure 3.1.

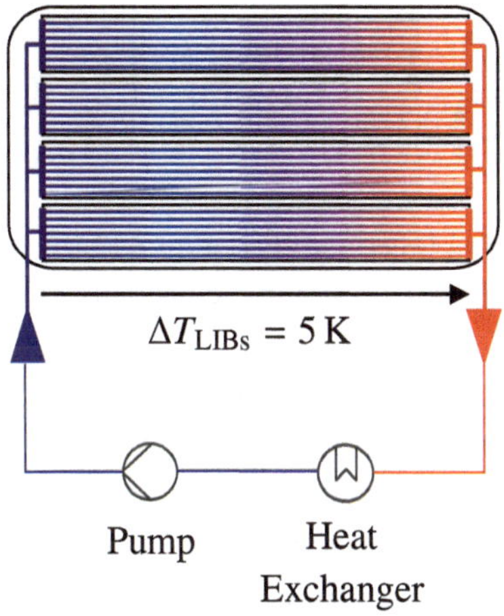

Figure 3.1: Schematic of a conventional reference BTMS architecture with unidirectional flow. Adapted from [71].

A conventional BTMS architecture consists of three key components, the battery pack cooling design, a heat exchanger and a pump. The heat exchanger is responsible for dissipating the LIB cell's waste heat to the envi-

M. Auch, *Holistic Modeling Framework for the Investigation of Innovative Battery Thermal Management Concepts*, Wissenschaftliche Reihe Fahrzeugtechnik Universität Stuttgart, https://doi.org/10.1007/978-3-658-52612-2_3

ronment/connected circuits and to ensure the required coolant battery pack inlet temperature. The pump delivers the required volume flow rate. In BEV packs, every heat sink and consequently, cooling channel is responsible to cool multiple LIB cells. As illustrated in Figure 3.2, the coolant temperature increases downstream in the cooling channel due to the absorption of waste heat from the LIB cells. As a result, the temperature difference between the coolant and the LIB cells decreases further along the flow path. This results in a reduced heat flow, leading to higher average temperatures for LIB cells located further downstream compared to those near the coolant inlet. Since this temperature imbalance cannot be completely eliminated, industry standards recommend limiting cell-to-cell temperature differences inside a BEV pack to $\Delta T_{\mathrm{LIB}} = 5\,\mathrm{K}$ [5].

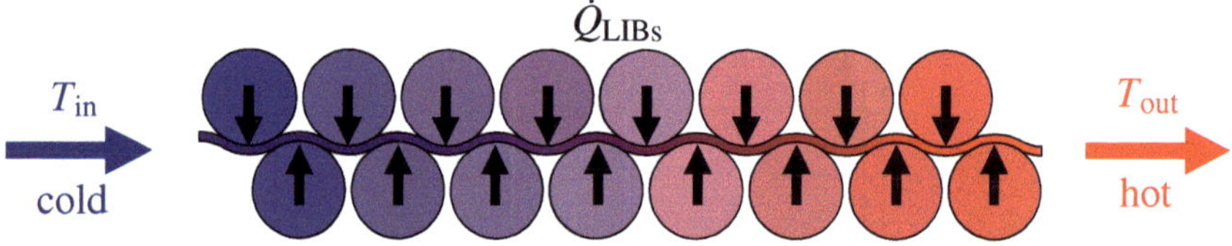

Figure 3.2: Influence of indirect liquid cooling BTMS concept coolant temperature increase on the operating temperature of the LIB cells.

However, as discussed in Section 2.3.2, the operating temperature directly impacts the degradation rate of LIB cells. Consequently, a cell-to-cell average temperature difference of 5 K can lead to temperature-related inhomogeneous aging inside the LIB modules and packs. As LIB pack performance is limited by the lowest capacity cell, this effect leads to a performance decrease over the BEV life-cycle [72]. This challenge is particularly critical in cylindrical LIB modules, where a large number of cells must be thermally managed within a limited space. In these systems, each heat sink is responsible for cooling multiple LIB cells, making uniform temperature distribution inherently difficult. As a result, cell-to-cell thermal imbalances remain a key bottleneck in current indirect liquid cooling BTMS architectures, necessitating novel solutions to enhance thermal homogeneity and battery longevity. Consequently, the first research question is formulated as: How can BTMS-related aging inhomogeneities within BEV LIB modules be reduced? To tackle this challenge, in this thesis a switchable thermal management BTMS concept is investigated to mitigate the cumulative temperature-related aging inhomogeneities inside LIB modules over the BEV life-cycle.

3.2 Switchable BTMS concept

The working principle and architecture of the proposed switchable BTMS are illustrated in Figure 3.3. In comparison to the conventional BTMS concept shown in Figure 3.1, the overall system architecture remains unchanged, with the sole modification being the integration of a 4/2 directional valve, as proposed in a recently granted patent [13]. This valve enables reversal of the coolant flow direction within the battery pack, effectively inverting the cell-to-cell temperature gradient. As a result, cells that were previously subjected to the highest temperatures become the coolest in subsequent load cycles and vice versa.

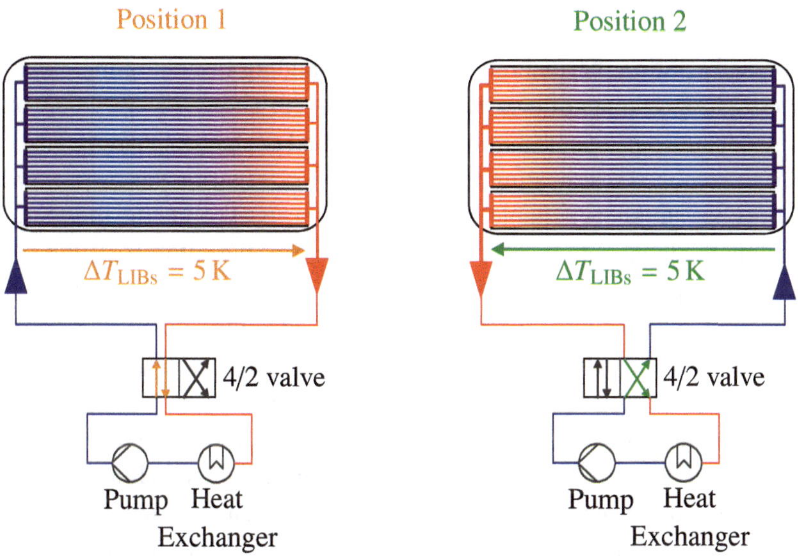

Figure 3.3: Schematic of the switchable BTMS architecture incorporating a 4/2 valve. Adapted from [71].

Moreover, considering current VTMS architectures, such as the Octovalve system introduced in Figure 2.2, the BTMS coolant inlet and outlet are already connected through a single centralized component. This configuration allows for easy integration of the additional 4/2 valve at the VTMS level, requiring only minimal architectural modifications.

While previous studies have explored the potential of altering the coolant flow direction, existing research has primarily focused on reciprocating flow approaches [12, 73, 74, 75]. The main objective of these approaches lies in actually reducing the occurring maximum temperature difference within the module during operation. For this purpose, the flow direction is reversed at fixed time intervals during for example one fast charging event, focusing on determining the optimum periodic time interval for switching, which requires either additional modifications to the BTMS architecture [74] or complex active control systems [12, 73, 75]. While this approach improves the temperature homogeneity within a module, the investigation and evaluation in these studies is limited to single charging or discharging events. Additionally, for indirect liquid cooling significant risks can be introduced, especially during thermal management demanding fast charging scenarios where a large amount of heat need to be removed. These risks include temporary reductions in heat absorption efficiency and coolant stagnation. This can lead to localized overheating and in worst-case scenarios thermal runaway.

In contrast, the switchable BTMS in this thesis expands the literature by tackling the temperature difference challenge over the entire BEV life-cycle. For this purpose, the proposed concept aims to optimize the operating strategy to mitigate cumulative, temperature-induced aging inhomogeneities that develop over the vehicle's lifetime, rather than improving the temperature uniformity during single charging/discharging processes. For this purpose, the switchable BTMS concept alters the flow direction only after each complete load cycle (charging and discharging). The switching process is exclusively performed during vehicle standstill, when no waste heat needs to be dissipated. This strategy ensures that the existing state-of-the-art BTMS architecture remains mainly unchanged, requiring only the addition of a 4/2 directional valve and a simple control strategy counting the load cycles for regulating the switching mechanism during vehicle standstill. The key distinctions between the existing literature and the proposed BTMS concept are summarized in Table 3.1.

Table 3.1: Comparison of existing research (reciprocating flow) and proposed switchable BTMS concept.

	Reciprocating flow strategies	Proposed switchable BTMS concept
Objective	Decrease of maximum temperature difference ΔT	Decrease of cumulative temperature-related aging inhomogeneities
Evaluation scope	Single event	Full life-cycle
Switching frequency	High (seconds,minutes)	Low, entire load cycle (hours,days)
Switching condition	During charging or driving	During vehicle standstill
System complexity	High (active control)	Low (load cycle count)
Risks	Coolant stagnation, overheating	-
Architecture changes	Moderate	Minimal (4/2-valve)

To quantitatively assess both the conventional unidirectional and switchable BTMS concepts in terms of their EOL characteristics, a module-level electro-thermal-fluid-aging model is required. This directly leads to the second research question: How must a reliable modeling workflow be defined to evaluate and investigate BTMS concepts with respect to their impact on the aging behavior of complete BEV battery packs?

3.3 Holistic thermal modeling approach

The objective of an accurate coupled electro-thermal-fluid-aging model at the full module level necessitates a 1D modeling approach due to the high computational effort required for life-cycle studies. While numerous studies address LIB modeling and optimization at the cell level, research at the full module level remains relatively limited [76]. Consequently, a holistic thermal modeling approach is developed in this thesis, as illustrated in Figure 3.4. It follows a hierarchical structure, progressing from the cell level to the module section and ultimately to the full module or pack level. On the cell level, an electrical ECM model, an aging model and 3D and 1D thermal models are required. The main characteristic of the module section level is including multiple LIB cells and considering the actual cooling design. In contrast to the cell level laboratory heat transfer environment, this enables investigating the influence of the applied cooling design on the predicted LIB cell temperatures in both 3D and 1D models. Finally, by scaling up the module section 1D model to the full module level and integrating the aging model, an approach for the life-cycle investigation of different BTMS concepts is derived.

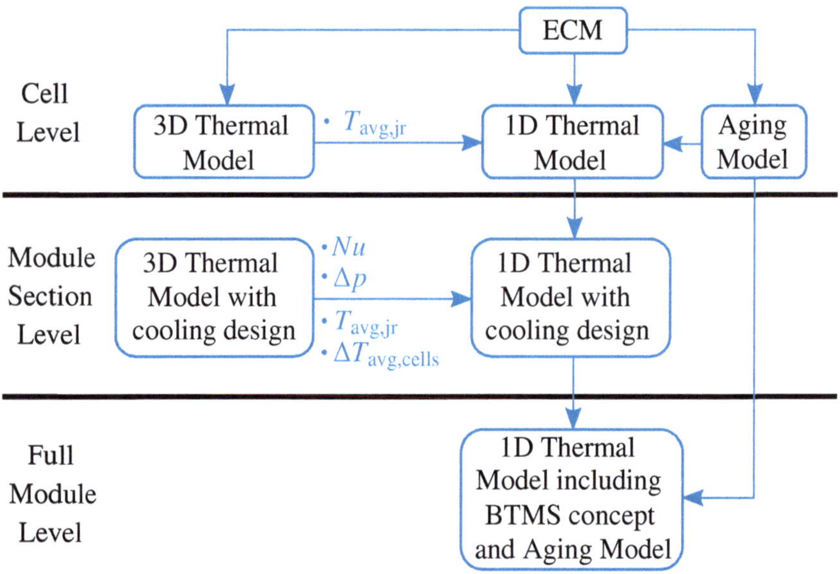

Figure 3.4: Development process of the full module-level 1D electro-thermal-fluid-aging model.

Several key aspects of this approach have to be highlighted. Starting on the cell level, there is a lack of 3D and 1D thermal cell models validated on multiple measurement positions over different load profiles. For this purpose, ECM-coupled thermal cylindrical cell models with a focus on bridging the gap between computational efficiency and accurate temperature prediction are developed.

Furthermore, the aging behavior is dependent on the active material temperature of the LIB cell. However, measuring the internal LIB temperature is a challenging task, because of accessing the interior of the LIB cell while ensuring not damaging the cell and sealing the inserted measurement sensor. Consequently, extensively validated LIB cell models represent a good option to access the internal jelly roll temperature. For this purpose, the average jelly roll temperature derived from detailed 3D thermal cell models can be used to compare and ensure the accuracy of the simplified 1D predicted temperature, which is the input parameter for the aging model. [77]

Finally, cooling designs in BEV packs are often customized, due to the requirements of minimal packaging space, optimum heat transfer to power demand characteristics and low cost. Consequently, heat transfer and pressure loss correlations are often not available for those specific designs. For this purpose, 3D computational fluid dynamics (CFD) tools are used to derive those correlations for the customized designs, which can be employed to 1D models. Additionally, the 3D module section level results can be used to ensure the accuracy of the 1D model with applied cooling design. Given the focus of this thesis on temperature-related aging inhomogeneities in LIB modules, special emphasis is placed on capturing and analyzing temperature differences across LIB cells.

Parts of this thesis have been previously published by the author. These include the influence of different ECM architectures and parameter dependencies on predicted voltage and heat generation [29], a simple, cost-effective and reliable experimental method for determining the specific heat capacity of cylindrical LIB cells [78], an improved ECM-coupled 3D thermal model for cylindrical LIB cells [79] and an evaluation of the potential benefits of a switchable compared to a unidirectional BTMS concept [71]. Shortly after the submission of the paper on the switchable BTMS concept [71], Talele et al. [80] published a study exploring a similar BTMS concept. However, their study is restricted to a small battery module comprising 15 LIB cells and a switching interval of 50 cycles. Furthermore, a different BTMS cooling design is used and their study is limited to 700 cycles, using a 3D simulation approach.

4 Modeling

To evaluate the potential life-cycle performance improvements enabled by the proposed switchable BTMS concept, a benchmark battery pack architecture must first be established. For this purpose, a specific LIB cell suitable for the pack design needs to be presented. This reference design serves as the foundation for the application of the holistic thermal modeling approach developed in this thesis. The subsequent modeling sections are organized according to the holistic modeling approach hierarchy, beginning on the cell to the module section and finally, to the full module level.

4.1 Reference battery pack

The cylindrical LIB cell LG21700-M50(T) is selected due to its current application in BEVs on the market (e.g. in the 2022 Tesla Model 3 Long Range variant E5LD battery pack [81]). This LIB cell features an NMC 811 cathode and a graphite-silicon anode, making it well-suited for high-performance BEVs. The cell characteristics, as specified in the manufacturer's data sheet [82], are summarized in Table 4.1.

Table 4.1: Electrical cell characteristics of the LG21700 M50(T) cell [82].

Name	LG INR21700 M50(T)
Energy in Wh	Nominal: 18.2 Minimum: 17.6
Nominal voltage in V	3.63
Voltage range in V	2.5 – 4.2
Standard charge	Constant current: 0.3 C Constant voltage 4.2 V End current (cut off) 50 mA
Standard discharge	Constant current 0.2 C End voltage (cut off) 2.5 V
Mass in g	70.0

M. Auch, *Holistic Modeling Framework for the Investigation of Innovative Battery Thermal Management Concepts*, Wissenschaftliche Reihe Fahrzeugtechnik Universität Stuttgart, https://doi.org/10.1007/978-3-658-52612-2_4

The reference battery pack and its BTMS cooling design are illustrated in Figure 4.1. To ensure a realistic evaluation of the studied BTMS concepts, the reference design is based on mid-class BEVs on the market [83]. The battery pack consists of 4416 cylindrical LIB cells in a 96s46p configuration. Considering the capacity and voltage of the studied LIB cell, this results in a battery pack capacity of approximately 76 kWh. In this thesis, the LIB cells are divided into four equally sized modules, each featuring a 24s46p cell arrangement. To ensure thermal symmetry of the LIB pack, the total battery coolant volume flow rate is evenly distributed among the four modules.

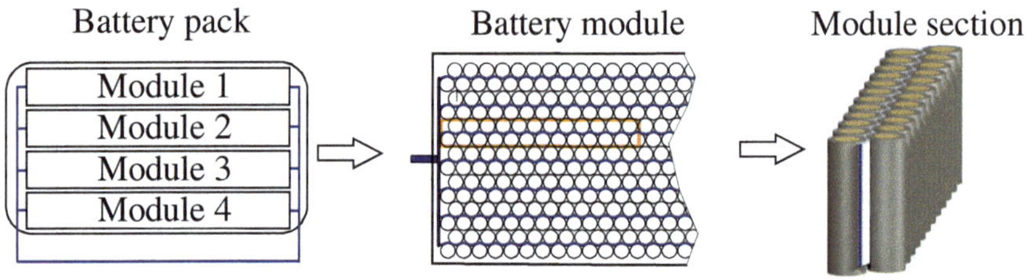

Figure 4.1: Schematic of reference battery pack, module and module section design including its thermal management system. Adapted from [71].

At the full module level, the cooling system consists of 14 linear arranged cylindrical LIB cell rows and seven stacked wavy mini-channel heat sinks. The battery module coolant flow rate is equally split among these seven heat sinks. Additionally, each stacked heat sink contains 30 mini-channels, further subdividing the heat sink coolant flow rate into individual channels to optimize heat dissipation. These stacked wavy mini-channel heat sink designs are often chosen, since they minimize packaging space while increasing the heat transfer surface area and improving the overall thermal performance of side-cooling BTMS approaches.

Having introduced the reference battery pack and BTMS design, the following sections present the developed modeling approaches, beginning at the cell level.

4.2 Cell level modeling

This section outlines the modeling approaches applied at the cell level, including the electrical model, the 3D and 1D thermal models and the aging model for the studied cylindrical LIB cell.

4.2.1 Electrical model

As introduced in Section 2.3.2, the electrical model is responsible for capturing the voltage behavior of LIB cells. For this purpose, the Dual Polarization ECM architecture is employed in this thesis. The ECM development process consists of two main parts, which are the experimental characterization and the ECM parameter identification. The detailed workflow and the testing protocols used have been previously published in [29]. Therefore, only the key aspects are summarized in this section.

Experimental characterization

To characterize the studied LIB cells, two different electrical test benches are used. Their main characteristics are presented in Table 4.2. Both test benches are manufactured by BasyTec and differ primarily in the number of available test channels and their current range. The BasyTec HPS system is a single-channel battery test bench used for all electro-thermal characterization experiments to avoid potential thermal behavior influences caused by multiple cells within the same chamber. The 20-channel BasyTec XCTS Mk 11 system is selected for the characterization of electrical cell-to-cell variations and the aging study. The measurement uncertainties for both test benches are provided in Table 4.2.

In both setups, temperature control is achieved using climate chambers manufactured by Weiss Umwelttechnik GmbH. In the experimental configuration, the LIB cells are mounted in cylindrical cell holders (FXB-CY-01-05) provided by Feinmetall. These holders are selected as the electrical contacting system due to their low contact resistance, which is enabled by a three-pin design. This design minimizes the influence of the contacting

Table 4.2: Key specifications of the BasyTec battery test benches used in electro-thermal and aging experiments.

Name	HPS.X.63.44.0595	XCTS Mk 11
Company	Basytec	Basytec
Maximum voltage in V	6	6
Voltage accuracy in mV	0.05 % F.S.	1
Maximum current in A	240/60	40
Current accuracy in A	0.05 % F.S.	0.01
Number of channels	1	20

system on the measured voltages and temperatures, thereby ensuring high measurement accuracy [84].

For the experimental characterization, four test procedures are required. These include capacity tests, OCV tests, a Hybrid Pulse Power Characterization (HPPC) test and an entropic coefficient test. For the capacity test, a constant ambient temperature is maintained using the climate chamber. The studied LIB is charged and discharged according to the nominal capacity Q_N specifications provided in the data sheet. Based on the measured experimental data, the cell capacity is determined using the ampere-hour integration method over the full discharge process. This test procedure can be repeated under varying temperatures and C-rates.

The OCV-test requires setting various SOC levels during both charging and discharging phases. At each SOC level, a rest period ($t_{rest} = 2\,h$) must be observed to allow the cell voltage to reach chemical and thermal equilibrium. This ensures that the condition for measuring the OCV, in particular that no load is applied, is fulfilled. At lower SOC levels ($< 20\,\%$) the required rest time may increase to up to 4 h to ensure equilibrium is achieved. In addition to SOC, the OCV is influenced by temperature and charging/discharging conditions.

The HPPC test, as proposed in [85], is essential for determining the resistance and capacitance parameters of the ECM. In this test procedure, a current pulse is applied to the LIB cell and the resulting voltage response is analyzed to derive the cell's dynamic behavior. The pulse duration is a critical parameter for determining ECM characteristics, particularly the internal resistance of the cell. To capture the full spectrum of polarization effects, including ohmic resistance, charge transfer resistance and diffusion-related resistance, a pulse duration of 10 s is recommended and adopted to all applied pulses [86]. Typically, a discharge pulse is chosen for characterization, followed by a corresponding charge pulse to restore the cell to its initial SOC. Since the dynamic voltage response depends on the applied load condition, current pulses with varying C-rates are used. These pulses are conducted at multiple SOC levels, temperatures and for charging/discharging conditions to capture a comprehensive set of dynamic behaviors.

Given that both the OCV and HPPC tests require measurements across a range of SOC levels, temperatures and cycling conditions, they are combined in the experimental procedure to reduce characterization time. An example of the recorded voltage and current profiles at one test temperature for the combined OCV/HPPC procedure is illustrated in Appendix Figure A.1.

Finally, an additional entropic coefficient test is conducted to determine the entropic coefficient $\partial U_{\mathrm{OCV}}/\partial T$ of the studied LIB cell. This parameter is necessary for calculating the reversible heat losses of the cell. The test procedure involves bringing the LIB cell to thermal and electrochemical equilibrium at a set temperature. The temperature is then varied and the cell is rested until it reaches a new equilibrium state, allowing the response of the OCV to temperature changes to be measured. This procedure is repeated for multiple SOC levels to capture the entropic behavior across the full SOC range.

ECM parameter identification

While the LIB cell's capacity and entropic coefficient can be directly obtained from the experimental tests, the data from the combined OCV/HPPC test must be post-processed to extract the ECM parameters. To achieve this, the ECM parameter identification procedure is divided into two parts and is illustrated for a representative load pulse in Figure 4.2. First, the OCV is determined and second, the remaining ECM parameters (resistances and capacitances) are identified. The OCV U_{OCV} is taken after the rest period requirement is fulfilled, corresponding to the cell voltage U_{cell} at time step t_{a} in Figure 4.2.

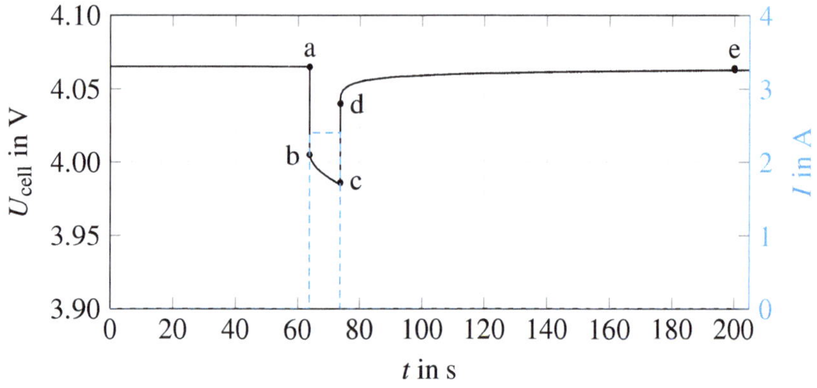

Figure 4.2: Measured voltage and current response to a 0.5 C discharge pulse used for parameter identification of the ECM. Adapted from [29].

Subsequently, the identification of the remaining parameters is performed using a slightly modified version of the method introduced by Zhu et al. [87]. In this approach, the relaxation voltage behavior captured during the experiment is used to derive the ECM parameters. The timestamps t_{a} and t_{c} are taken to determine the duration of the applied current pulse. For all current pulses, the end of the relaxation period is defined as $t_{\mathrm{e}} = t_{\mathrm{c}} + 600\,\mathrm{s}$ (position of t_{e} only exemplarily set in Figure 4.2). Since the OCV has already been determined, the voltage profile of each current pulse can be shifted so that the cell voltage at the end of the relaxation period is set to zero, resulting in $U_{\mathrm{cell}}(t_{\mathrm{e}}) = 0\,\mathrm{V}$.

This approach simplifies the parameter identification equation for the Dual Polarization Model to

$$U_{\mathrm{model}}(t) = -U_{\mathrm{RC,1}}\,\exp(-\frac{t - t_{\mathrm{c}}}{\tau_1}) - U_{\mathrm{RC,2}}\,\exp(-\frac{t - t_{\mathrm{c}}}{\tau_2}) - I R_0 \qquad (4.1)$$

where U_{model} represents the voltage predicted by the ECM. This equation is used to optimize the parameters $U_{\text{RC},i}$, τ_i and R_0 using MATLAB's gradient-based optimization function *lsqnonlin*, based on the measured voltage response in the time interval from t_c to t_e. Once the optimal parameters are obtained, the polarization resistances R_i can be calculated by

$$R_i = \frac{U_{\text{RC},i}}{I\left(1 - \exp(-\frac{t_\text{c}-t_\text{a}}{\tau_i})\right)} \tag{4.2}$$

for $i = 1, 2$. The outcome of the optimization process depends on the constraints imposed on each RC network parameter, particularly the time constants and corresponding capacitances. This parameter identification procedure is applied across all parameter combinations tested during the experimental characterization.

For the Dual Polarization ECM developed in this thesis, temperatures of $T = 5, 25, 45\,°\text{C}$, current loads of $I = 0.2, 0.5, 1, 2\,\text{C}$ with the C-rate related to the nominal capacity, 5 %-SOC steps and charging/discharging conditions have been integrated as parameter dependencies. Since this work focuses on the holistic thermal modeling workflow, only the temperature-dependent ECM parameters in Figure 4.3 are shown here, while all other parameter dependencies are provided in Appendix Section B. However, it is essential to confirm that no significant cell-to-cell variation exists for the studied LIB cell type. Therefore, five additional cells have been characterized and their corresponding ECM parameters have been identified. The results are presented in Appendix Section C, with the key findings summarized in the following.

The capacity variation among the five tested cells is less than 1 % relative to their arithmetic mean value of $Q_\text{N} = 4.74\,\text{Ah}$, which is used as the nominal capacity of the studied LIB cell throughout this thesis. Similarly, the deviation in internal direct current resistance is below 2.4 % for all cells compared to the mean value. For the selected Dual Polarization ECM, the deviations in OCV and ohmic resistance R_0 are both below 1.4 % relative to the arithmetic mean. However, more notable variations are observed in the RC network parameters, as shown in Appendix Table C.4. These deviations are attributed to the optimization algorithm used in the parameter identifi-

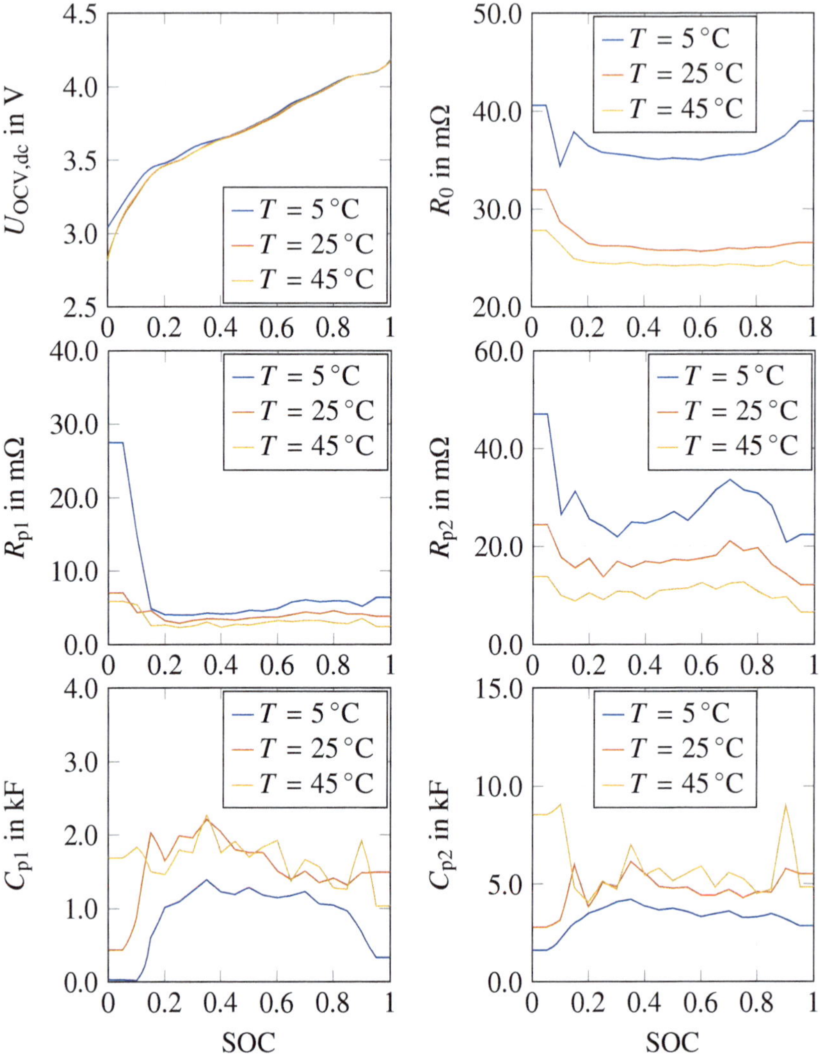

Figure 4.3: Derived ECM parameters for the single cell at three temperatures $T = 5, 25, 45\,°C$, determined using a 0.5 C discharge pulse.

cation process, which distributes the RC elements differently depending on the optimal fit to the measured data.

Although these variations are not considered significant, to minimize the potential influence of cell-to-cell variation, arithmetic mean values from the five tested cells are used wherever available. For the two temperature conditions at $T = 5, 45\,^{\circ}\mathrm{C}$ the single-cell ECM parameters are used due to limited data. Based on the identified ECM parameters, Equation (2.7) is solved to accurately capture the transient voltage behavior under various load cycles. In addition, by using the derived entropic coefficient, Equation (2.10) is employed to calculate the total heat generation of the studied LIB cell across its operating range. As both the ECM parameters and equations depend on the LIB cell temperature, the thermal modeling approaches developed in this thesis are presented in the following sections.

4.2.2 3D thermal model

As outlined in Section 2.3.3, 3D thermal models can be categorized based on the level of detail applied to the jelly roll and to the remaining cell components. This thesis introduces an improved ECM-coupled 3D thermal modeling approach, the multi-partition heat generation and thermal resistance (MPH-TR) model. The following section presents the MPH-TR model setup, including thermal properties and thermal resistances, which builds on the multi-partition approach introduced by Yang et al. [54]. For a more comprehensive description of the developed MPH-TR model, the reader is referred to a previous publication by the author [79].

Model setup

As illustrated in Figure 4.4, the MPH-TR approach is defined by two key characteristics. First, the use of a multi-partition heat generation strategy and second, the integration of the internal electrical insulation layers of the LIB cell as thermal resistances.

The MPH-TR model consists of three regions, the jelly roll, the positive terminal and the casing, as illustrated in Figure 4.4 b). The jelly roll is represented as a single region based on the assumption of homogeneous

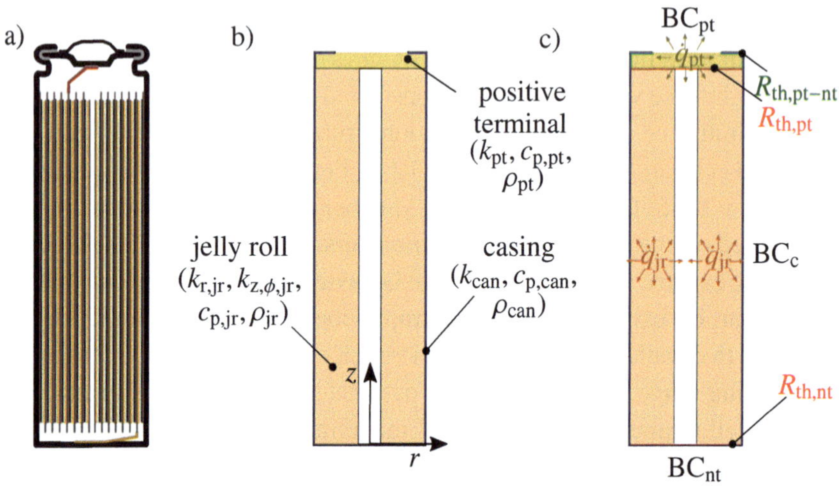

Figure 4.4: a) Schematic of a cylindrical LIB cell; b) Schematic of the 3D MPH-TR model; c) 3D MPH-TR model with applied boundary conditions. Adapted from [79].

heat generation and the use of effective thermal properties. On the remaining component side, the model considers both the casing and the positive terminal. The positive terminal combines the positive tab and cap into a single region, thereby reducing the discretization effort while preserving an accurate prediction of heat flow within the LIB cell. The negative terminal is not modeled as a separate region but is instead incorporated into the jelly roll region.

The total generated heat by the LIB cell is calculated using the Bernardi Equation (2.10). The derived ECM parameters and entropic coefficient of the studied LIB cell enable the calculation of both irreversible and reversible heat losses. To implement the multi-partition heat generation approach, the Bernardi equation is divided into two separate heat source components, which are the generated heat of the jelly roll $\dot{Q}_{jr}$ and of the positive terminal $\dot{Q}_{pt}$. This reduces Equation (2.12) to

$$\dot{Q}_{tot} = \dot{Q}_{jr} + \dot{Q}_{pt} \tag{4.3}$$

where the generated heat of the jelly roll, consisting of irreversible and reversible terms, is defined as

$$\dot{Q}_{\text{jr}} = R_{\text{el,jr}}\, I^2 + I\,T\,\frac{\partial U_{\text{OCV}}}{\partial T} \tag{4.4}$$

and the generated heat of the positive terminal, consisting of irreversible heat, is defined as

$$\dot{Q}_{\text{pt}} = R_{\text{el,pt}}\, I^2 \tag{4.5}$$

dependent on the positive terminal electrical resistance $R_{\text{el,pt}}$. By dividing the heat source terms by the corresponding component volume, the volumetric heat source can be obtained as $\dot{q}_i = \dot{Q}_i/V_i$, for each component i, as illustrated in Figure 4.4 c). To complete the MPH-TR modeling approach, the geometric dimensions of the studied LIB cell and the electrical resistance of the positive terminal must be determined. For this purpose, one sample of the studied LIB cell is disassembled, providing detailed insights into the cell's internal structure, as shown in Figure 4.5.

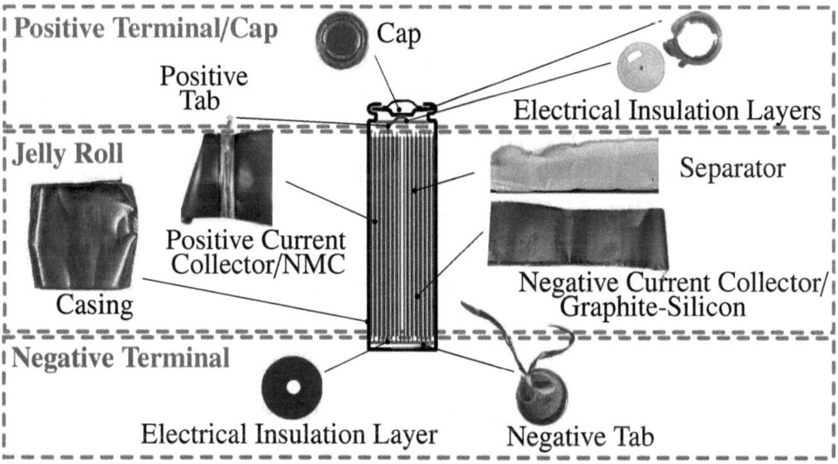

Figure 4.5: Disassembled LG21700 M50(T) LIB cell. Identified components grouped into positive terminal/cap, jelly roll and negative terminal.

The disassembled components can be categorized into three different groups, the positive terminal, the jelly roll and the negative terminal. The positive terminal group consists of the LIB cap, the positive tab and two layers of electrical insulation material. These insulation layers are used to electrically isolate the positive and negative terminals, as well as the current collectors, to prevent potential short circuits. The jelly roll comprises the previously described layered structure and features a single-tab design for both the positive and negative tabs. The negative terminal group includes an electrical insulation layer and the negative tab. Finally, the entire cell is enclosed in a cylindrical casing with a wall thickness of 0.25 mm. The disassembly procedure is used to determine the geometric parameters of the defined MPH-TR regions, which are summarized in Appendix Table D.1.

Using the disassembled components, the electrical resistance of the positive terminal can be determined. The disassembly revealed that the positive tab is wrapped in Kapton tape, which electrically isolates it from all other components within the jelly roll, except for the positive current collector. Due to the low thermal conductivity of Kapton tape ($k = 0.12-0.2\,{}^{\mathrm{W}}/_{\mathrm{mK}}$), this wrapping also provides thermal isolation, reducing heat transfer between the positive tab and the surrounding components within the jelly roll. The connection between the positive tab and the LIB cap is established by soldering, while a conductive adhesive is used between the positive current collector and the positive tab. These adhesives are typically characterized by high electrical conductivity but low thermal conductivity ($k = 0.5 - 3\,{}^{\mathrm{W}}/_{\mathrm{mK}}$). As a result, the MPH-TR approach assumes that the positive tab generated heat is transferred entirely to the positive terminal of the LIB cell. Consequently, the electrical resistance of the positive terminal $R_{\mathrm{el,pt}}$ is defined as

$$R_{\mathrm{el,pt}} = R_{\mathrm{el,cap}} + R_{\mathrm{el,ptab}} \tag{4.6}$$

where $R_{\mathrm{el,cap}}$ and $R_{\mathrm{el,ptab}}$ represent the electrical resistances of the cap and the positive tab, respectively. In the previous study by the author [79], the electrical resistances of the cap, the positive and negative tab were determined. The corresponding results are presented in Table 4.3.

The positive tab alone is the primary contributor to the electrical resistance of the positive terminal, accounting for 18 % of the total internal resistance of the LIB cell within the 20-100 % SOC range. Together, the positive tab and cap contribute for 20.41 % of the total internal resistance of the LIB cell

Table 4.3: Electrical resistance of the LIB cap, positive and negative tab (*m*=measured, *c*=calculated). Out of [79].

	Material	Parameter	R_{el} in mΩ
Cap	Aluminum/steel	$R_{\mathrm{el,cap}}$	0.85^{m}
Positive tab	Aluminum	$R_{\mathrm{el,ptab}}$	5.68^{c}
Negative tab	Copper	$R_{\mathrm{el,ntab}}$	0.91^{c}

and consequently to the overall generated heat. Considering the relatively small volume and mass of the cap compared to the other components, this explains the frequently observed highest temperatures at the positive terminal during high C-rate operation. In contrast, the negative tab, which is made of copper with significantly lower electrical resistivity compared to the positive tab material aluminum, exhibits an electrical resistance almost an order of magnitude smaller. Consequently, the heat generated by the negative tab is not treated as a separate heat source. Instead, it is assigned to the jelly roll heat generation due to its low electrical resistance and its efficient thermal coupling to both the cell casing and the jelly roll. These findings conclude in the previously introduced model-setup of the MPH-TR approach, as illustrated in Figure 4.4 c).

Thermal properties and resistances

To apply the MPH-TR model to the studied LIB cell, the effective thermal properties density ρ, specific heat capacity c_{p} and thermal conductivity k need to be determined. In addition, the identified electrical insulation layers are modeled as thermal resistances R_{th} and applied at the corresponding region interfaces, as illustrated in Figure 4.4 c). This modeling approach reduces discretization and computational effort while maintaining the key thermal characteristics associated with the internal structure of the studied LIB cell.

To determine the densities of the corresponding regions, the masses of the disassembled components jelly roll, positive terminal and casing are measured. By calculating the respective volumes defined in the MPH-TR model, the density for each region can be derived. This approach ensures that the thermal behavior is accurately captured, even though the detailed internal structures of the positive terminal and jelly roll are simplified. The resulting density values are presented in Table 4.4.

Table 4.4: Mass m, volume V and density ρ of the whole LIB cell and the defined regions in the 3D MPH-TR model (m=measured, c=calculated).

Region	m in g	V in m^3	ρ in kg/m^3
Whole cell	68.423^m	$2.4529 \cdot 10^{-5}$	2789.5^c
Jelly roll	57.781^c	$2.1498 \cdot 10^{-5}$	2687.7^c
Positive terminal	1.306^m	$9.5095 \cdot 10^{-7}$	1399.2^c
Casing	9.345^m	$1.3113 \cdot 10^{-6}$	7126.5^c

Since the specific heat capacity of the LIB cell is a critical parameter for the accuracy of thermal models, a novel in-situ measurement method was developed as part of this thesis. Details on the measurement procedure and its validation are provided in a previous publication [78]. The LIB jelly roll exhibits anisotropic thermal conductivity due to the layered structure, which results in different heat flow paths depending on whether heat propagates serially or in parallel through the layers. This anisotropy significantly influences the heat transfer behavior, particularly with respect to various potential BTMS cooling configurations (bottom, top or side cooling). Consequently, the jelly roll thermal conductivity is separated into radial $k_{r,jr}$ and axial/circumferential $k_{z,\phi,jr}$ components. In contrast, the casing and positive terminal (each composed of a single solid material) are assumed to have isotropic thermal conductivity. The values for specific heat capacity and thermal conductivity used in the MPH-TR approach for the studied cell are summarized in Table 4.5.

Finally, the identified electrical insulation layers are modeled in a simplified manner as thermal resistances between regions, as indicated in Figure 4.4 c). For this purpose, the thermal conductivity of the electrical insulation material is set to $k = 0.1$ W/mK. Using the measured thicknesses of the insulation layers obtained from the cell disassembly, the thermal resistances R_{th}

Table 4.5: Specific heat capacity c_p and thermal conductivity k of the whole LIB cell and the defined regions in the 3D MPH-TR model (c=calculated, e=estimated, [ ]=literature, T in °C).

Region	c_p in J/kgK	k_r in W/mK	$k_{z,\phi}$ in W/mK
Whole cell	$1.8\,T + 774.93$ [78]	$1.39-0.0059\,(T-26.85)^c$	30 [88]
Jelly roll	$1.8\,T + 813.51^c$	$1.35-0.0059\,(T-26.85)$ [89]	10^e
Positive terminal	565.4 [55]	30 [55]	30 [55]
Casing	456 [90]	20 [91]	20 [91]

are calculated with Equation (2.14). The resulting values are summarized in Table 4.6. The transferred heat between the regions is proportionally influenced by the insulation layer thickness. Consequently, the thermal coupling between the positive terminal and the jelly roll decreases the most, whereas the heat transfer between the negative terminal and the jelly roll remains more than five times higher.

Table 4.6: Applied thermal resistances in the 3D MPH-TR model (m=measured, c=calculated).

Interface	Parameter	Insulation thickness in mm	R_{th} in m²K/W
PT-Jelly roll	$R_{th,pt}$	1.10^m	0.0110^c
NT-Jelly roll	$R_{th,nt}$	0.20^m	0.0020^c
NT-PT	$R_{th,pt-nt}$	0.65^m	0.0065^c

As illustrated in Figure 4.4 c), the boundary conditions (BCs) differ between the corresponding LIB cell outer surfaces (cell mantle, positive terminal and negative terminal). Depending on the experimental setup under consideration, the heat transfer boundary conditions can be adjusted accordingly and evaluated at each time step based on the corresponding surface-averaged and ambient temperatures. The jelly roll and positive terminal regions are discretized using a polyhedral mesh, while the thin mesher is used for the casing, resulting in a minimum of three cells in each spatial direction to ensure accurate resolution of heat conduction effects. A grid convergence study was conducted to confirm that the discretization does not influence the simulation results [79].

4.2.3 1D thermal model

Building upon the insights gained from the development of the MPH-TR 3D thermal model and the cell disassembly, a thermal resistance network has been selected as the foundation for the 1D modeling approach. Similar to the 3D model, this represents a modified implementation of the multi-partition concept originally proposed by Yang et al. [54].

Model setup

As shown in Figure 4.6, the primary distinction from the 3D model lies in the abstraction of the LIB cell into a network of seven lumped thermal mass nodes. These thermal nodes represent the following regions of the LIB cell, which are (1) mean jelly roll, (2) top jelly roll, (3) bottom jelly roll, (4) positive terminal, (5) negative terminal, (6) casing mantle on one side and (7) casing mantle on the opposite side.

The mean jelly roll thermal mass node (1) plays a critical role in the 1D thermal model, as it provides the temperature input for the ECM (which is illustrated in Figure 4.10). The additional jelly roll nodes (2 and 3) are included to capture axial temperature gradients along the jelly roll. The positive terminal node (4) is essential for implementing the multi-partition heat generation approach. Consistent with the 3D MPH-TR model, the total heat generation $\dot{Q}_{\mathrm{tot}}$ from the ECM is partitioned into the positive terminal (4) $\dot{Q}_{\mathrm{pt}}$ and the mean jelly roll (1) $\dot{Q}_{\mathrm{jr}}$. Furthermore, both the positive (4) and negative (5) terminal nodes are required to investigate top and bottom cooling strategies. Finally, the casing mantle nodes (6 and 7) are incorporated to enable the evaluation of one-sided and two-sided side-cooling approaches.

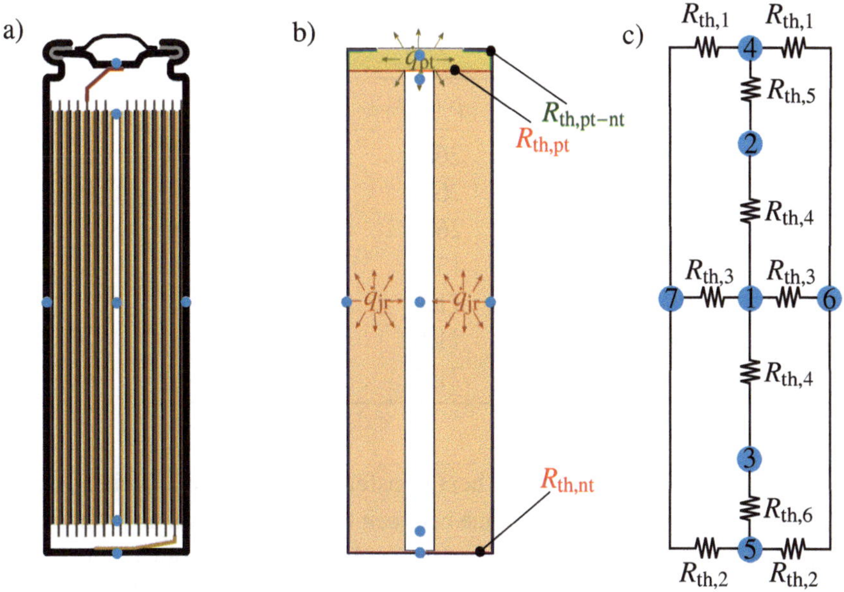

Figure 4.6: a) General structure of a cylindrical LIB cell with thermal network nodes in light blue; b) 3D MPH-TR model with thermal network nodes in light blue; c) Developed 1D thermal resistance network model. Adapted from [71].

Thermal properties and resistances

Based on the measured region masses from the MPH-TR model provided in Table 4.4, the corresponding component mass is assigned to each thermal mass node, as detailed in Table 4.7. To avoid significant influence on the mean jelly roll temperature prediction, the two additional jelly roll nodes (2 and 3) are each assigned a minimum mass of 1 g. The total casing mass is distributed between the negative terminal node (5) and the two casing mantle nodes (6 and 7) represented in the 1D model.

Since each node simplifies a 3D component to a point mass, heat conduction between the thermal mass nodes is modeled using the thermal resistance network, as illustrated in Figure 4.6 c). For this purpose, Equation (2.14) is

Table 4.7: Mass m, density ρ and specific heat capacity c_p used in the developed 1D thermal model for each node (T in °C).

Node	m in g	ρ in $^{kg}/_{m^3}$	c_p in $^{J}/_{kg\,K}$
(1)	55.78	2688	$1.8\,T + 813.51$
(2)	1.00	2688	$1.8\,T + 813.51$
(3)	1.00	2688	$1.8\,T + 813.51$
(4)	1.31	1399	566
(5)	2.00	7127	456
(6)	3.67	7127	456
(7)	3.67	7127	456

applied using the corresponding heat transfer surface A, the associated thermal conductivity k and the distance between the thermal mass nodes derived from the 3D model. The thermal resistances $R_{th,1} = 67.0\,^{K}/_{W}$, $R_{th,5} = 33.0\,^{K}/_{W}$ and $R_{th,6} = 6.0\,^{K}/_{W}$ primarily result from the identified electrical insulation layers integrated within the LIB cell. Heat conduction through the casing contributes a thermal resistance of $R_{th,2} = 274.9\,^{K}/_{W}$, which is determined by the casing geometry and its thermal conductivity. The resistance between the jelly roll nodes $R_{th,4} = 10.0\,^{K}/_{W}$ is based on the axial thermal conductivity of the jelly roll. Finally, heat transfer through a hollow cylinder is described by

$$\dot{Q} = \frac{2\pi\, l\, k_{r,jr}}{\ln(\frac{r_o}{r_i})}(T_i - T_o) \tag{4.7}$$

and is used to calculate the thermal resistance $R_{th,3}$, representing radial conduction from the mean jelly roll node (1) to each casing mantle node (6 and 7). The radial thermal conductivity $k_{r,jr}$ is temperature-dependent, as given in Table 4.5. The outer radius is determined from the measured cell geometry, while the inner radius is defined as $r_i = 7.5$ mm, thereby dividing the jelly roll into two equal volumes. This approach ensures that the mean jelly roll node (1) accurately represents the volume-averaged jelly roll temperature.

The thermal mass nodes at the top (4), bottom (5) and casing mantle (6 and 7) include additional thermal connections that vary depending on the applied boundary conditions and the specific installation configuration of the cells within the battery pack. The developed 1D ECM-coupled thermal model enables the consideration of temperature-dependent aging behavior of the studied LIB cell, while maintaining low computational cost. The following section introduces the corresponding aging modeling approach.

4.2.4 Aging model

The aging model is required to capture macroscopic effects, which are both capacity fade and internal resistance growth over the life-cycle of the studied LIB cell. For this purpose, the percentage of cycling-related capacity loss Q_{loss}^{cyc} can be calculated using an Arrhenius based degradation model proposed by Wang et al. [92], which depends on the ampere-hour throughput Ah of the LIB cell

$$Q_{\text{loss}}^{\text{cyc}} = B_{\text{cyc}} \exp(\frac{-E_{\text{a,cyc}} + \alpha \, |I|}{RT}) \, Ah^{z_{\text{cyc}}} \tag{4.8}$$

where B_{cyc} represents a pre-exponential factor, $E_{\text{a,cyc}}$ is the activation energy, α represents a factor for the current-dependent influence on the activation energy, I is the applied current, R represents the universal gas constant, T is the average LIB cell temperature, Ah is the total charge passed through the LIB cell and z_{cyc} represents the power law factor. The degradation behavior depends on z_{cyc}, determining whether it is sublinear (< 1), linear ($= 1$) or superlinear (> 1).

The proposed modeling approach is valid for specific temperature ranges, since different dominant aging effects are present [6, 92]. As a consequence, the sign of the activation energy $E_{\text{a,cyc}}$ depends on the modeling of aging effects above $T > T_{\text{optimum}}$ or below $T < T_{\text{optimum}}$ the optimum operating temperature [6]. For lower temperatures, lithium-plating is the main degradation effect, while for higher temperatures SEI-layer growth is assumed to be the predominately degradation effect. Furthermore, the optimum operating temperature can vary dependent on the cell chemistry, manufacturer, type and C-rate [42, 93].

Fleckenstein et al. [45] demonstrated that internal temperature gradients within a LIB cell can significantly affect macroscopic cell-level parameters and proposed considering this effect by increasing the average temperature by 10 % of the maximum temperature difference within the cell. However, recent work by Paarmann et al. [94] suggests, that cell-level thermal gradients primarily influence the local electrode scale and do not significantly change the macroscopic parameters of capacity fade and resistance growth [94]. Consequently, this thesis assumes that using the average cell temperature is adequate for capturing the dominant temperature-related aging effects of the studied LIB cell.

Although there is no direct physical relationship, the measured increase in resistance R_{gain}^{cyc} can empirically be correlated to the measured capacity loss using

$$R_{\text{gain}}^{cyc} = \alpha_{\text{cyc}}^{R} \, (Q_{\text{loss}}^{\text{cyc}})^{z_{\text{cyc}}^{R}} \tag{4.9}$$

where α_{cyc}^{R} represents a pre-exponential factor and z_{cyc}^{R} presents a dimensionless constant. Having introduced all cell level modeling approaches, in the following the module section level is described.

4.3 Module section level modeling

While accurate heat flow prediction within the cell is the primary objective at the LIB cell level, thermal modeling at the module or pack level requires additional consideration of heat transfer to the heat sink. This emphasizes the critical role of the battery pack and BTMS design. To address this, a module section comprising 24 LIB cells is extracted from the full module-level reference BTMS design, as introduced in Section 4.1. In this thesis, both 3D and 1D models are developed for the module section level, with particular emphasis on using the 3D model to ensure the accuracy of the 1D model in capturing cell-to-cell temperature variations among neighboring cells.

4.3.1 3D thermal model

The 3D module section model includes 24 LIB cells (in a 3s8p configuration), the positive and negative busbars (copper, positioned on top), the heat sink (aluminum) and the structural foam (polyurethane).

Model setup

The model setup for the module section is illustrated in Figure 4.7. The design features a dual-row linear arrangement of cylindrical cells, with a central wavy mini-channel heat sink positioned between the two rows. The cell-to-cell spacing within the linear arrangement is set to 1.5 mm to reduce the risk of thermal propagation [95].

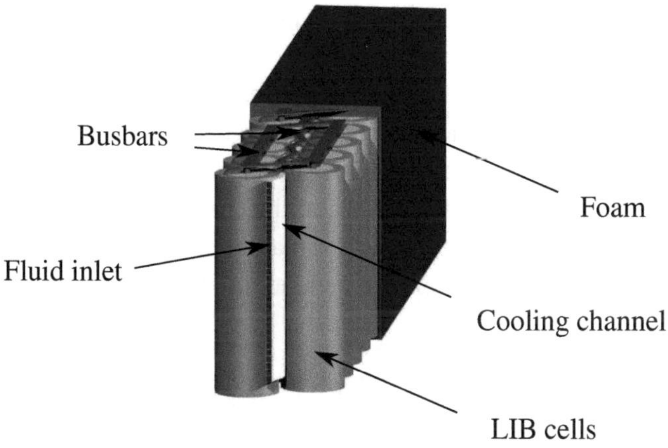

Figure 4.7: Illustration of the 3D geometry of the analyzed battery module section. Adapted from [71].

At both the front and rear ends of the module section, the foam is aligned with the inlet and outlet of the cooling channel. Furthermore, a minimum foam thickness of 1 mm is maintained between the cells and the left and right surfaces of the foam. The design also incorporates a double-top busbar configuration, recently patented by Tesla [96], which facilitates the potential integration of cell bottom venting systems for improved thermal runaway management.

Thermal properties and resistances

All 24 LIB cells are modeled using the MPH-TR model presented in Section 4.2.2. The thermal properties of the additional materials are documented in Appendix Table D.2. Ethylene glycol mixed with water in a 50:50 volume ratio is employed as the cooling medium, selected for its lowered freezing point and favorable heat transfer characteristics. The temperature-dependent thermophysical properties of this coolant mixture, as used in the simulations, are provided in Appendix Table D.3.

As illustrated in Figure 4.8, periodic boundary conditions are applied to the left and right sidewalls of the foam to realistically represent the integration of the module section into the complete battery module, where neighboring cells would be present. Due to the electrical conductivity of aluminum, which is used for the cooling channel heat sink, an additional layer of electrical insulation is required to prevent current flow between the LIB cell and the cooling structure. For this purpose, an insulation material with a thickness of 0.1 mm and a thermal conductivity of 0.1 $^{\mathrm{W}}/_{\mathrm{m\,K}}$ is assumed. This insulation layer is modeled as a thermal resistance of $R_{\mathrm{th,hs}} = 0.001$ $^{\mathrm{m^2\,K}}/_{\mathrm{W}}$ and is applied over the entire outer surface of the cooling channel heat sink. A minimum of three computational cells is maintained in all spatial directions to ensure accurate resolution of heat conduction within solid components.

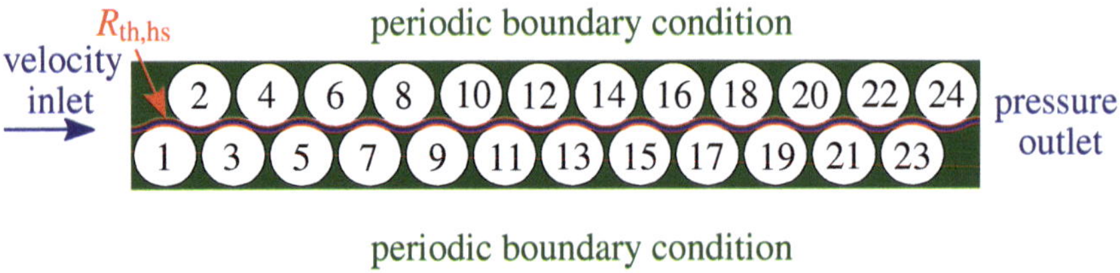

Figure 4.8: Boundary conditions and cell numbering of the studied 3D battery module section.

Special emphasis is placed on the fluid region of the 3D model, as it is critical for accurately capturing heat dissipation from the LIB cells and consequently, the overall thermal behavior of the battery module section under the applied cooling design. The fluid boundary conditions are defined by a velocity inlet and a pressure outlet. The detailed modeling approach for the CFD-based mini-channel heat sink is introduced in the following section.

Heat sink

To evaluate the performance of the heat sink, three different mini-channel designs are analyzed. As shown in Figure 4.9, the investigated configurations include a single straight channel, a single wavy channel and a stacked wavy mini-channel (consistent with the design used in the reference BTMS), all featuring a rectangular cross-section. The straight channel design is included to validate the heat transfer characteristics obtained from the CFD simulations, as there are no established correlations in the literature for wavy channels with rectangular cross-sections. The key geometric parameters, including the cross-section dimensions, wavy amplitude A and wavelength λ, are summarized in Table 4.8.

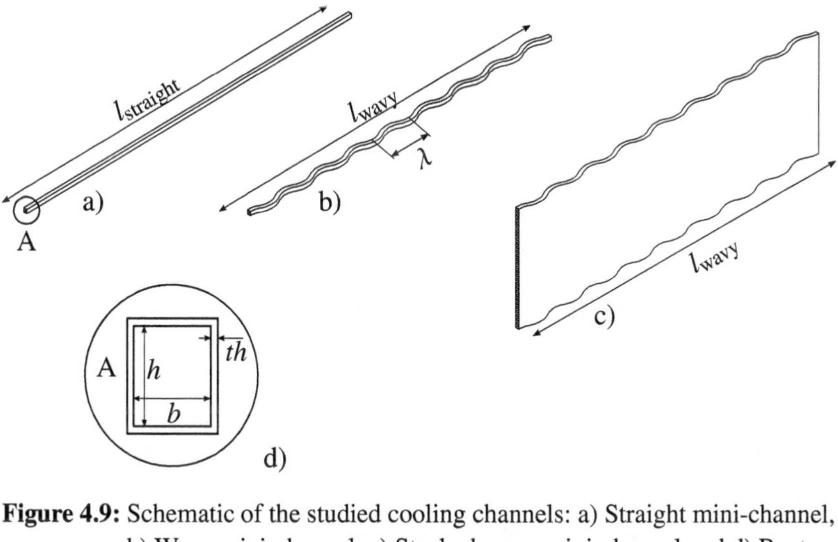

Figure 4.9: Schematic of the studied cooling channels: a) Straight mini-channel, b) Wavy mini-channel, c) Stacked wavy mini-channel and d) Rectangular channel cross-section. Adapted from [71].

Table 4.8: Geometrical parameters of the (wavy) mini-channels.

Parameter	Value
Channel height h in mm	2.00
Channel width b in mm	1.70
Hydraulic diameter d_h in mm	1.84
Rectangular aspect ratio $^b/_h$	0.85
Cooling fin thickness th in mm	0.15
Wavy channel amplitude A in mm	1.48
Wavy channel wavelength λ in mm	22.60
Straight channel length l_straight in m	1.60
Wavy channel length l_wavy in m	0.52
Number of channels in stack n_stack	30

The wavy mini-channel fluid domain is discretized using a structured, directed mesh incorporating prism layers to ensure accurate resolution of the boundary layer. Special emphasis is placed on the prism layer discretization in the rectangular corners of the fluid region to maintain an appropriate cell growth rate perpendicular to the channel walls. Furthermore, for accurate determination of local heat transfer characteristics in laminar mini-channel flow, mesh resolution along the channel length is particularly important to capture axial variations in temperature and flow properties. In the commercial CFD solver STAR-CCM+, a directed structured mesh cannot be conformal with a separately defined part. Therefore, a mapped contact interface between the cooling channel wall and the fluid domain is employed to accurately capture the heat transfer characteristics at the fluid-solid interface.

Finally, a grid convergence study is successfully conducted, with the results being shown in Appendix Table D.4. The study reveals that, when using the fine mesh, the predicted heat transfer and pressure loss are not significantly affected by further mesh refinement. Consequently, the fine mesh illustrated in Appendix Figure D.1 is selected for all subsequent heat sink simulations.

4.3.2 1D thermal model

The 1D thermal model of the module section includes the same components as those incorporated in the 3D model, with the exception of the busbars. In the 1D model, the busbars are neglected, as it is assumed that they primarily contribute to local temperature inhomogeneities rather than to a significant cell-to-cell temperature rise between the LIB cells from front to rear within the module.

Model setup

The schematic representation of the 1D model for the 24-cell module section is illustrated exemplarily for two cells in Figure 4.10. Compared to the cell-level 1D model, the module section model incorporates the two additional thermal mass nodes cooling channel (8) and foam (9). Due to the one-sided side-cooling design of the reference BTMS concept, the cooling channel (8) is thermally connected to only one of the two LIB cell casing nodes (6), enabling accurate calculation of the temperature gradients and heat flow between coolant and mean jelly roll. The remaining outer LIB cell nodes, consisting of the opposite casing side (7), the positive terminal (4) and the negative terminal (5), are thermally connected to the structural foam node (9). This configuration enables accurate prediction of temperature gradients and heat flow between the LIB cell and the coolant. The coolant itself is coupled to the cooling channel thermal mass node (8) via a thermal-hydraulic internal flow module with heat exchange (10).

Thermal properties and resistances

The integration of the module section level components requires the determination for the thermal resistances between the newly added thermal mass nodes. Consistently to the LIB cell model, the 3D model is used to derive the associated heat transfer surface A, thermal conductivity k and distance between the corresponding thermal mass nodes. The insulation layer between the cooling channel and the LIB cell casing/foam mainly contributes to the resulting thermal resistance of $R_{th,7} = 1.3\,{}^{K}/_{W}$. Due to the low thermal conductivity ($0.03\,{}^{W}/_{mK}$) of the polyurethane-based structural

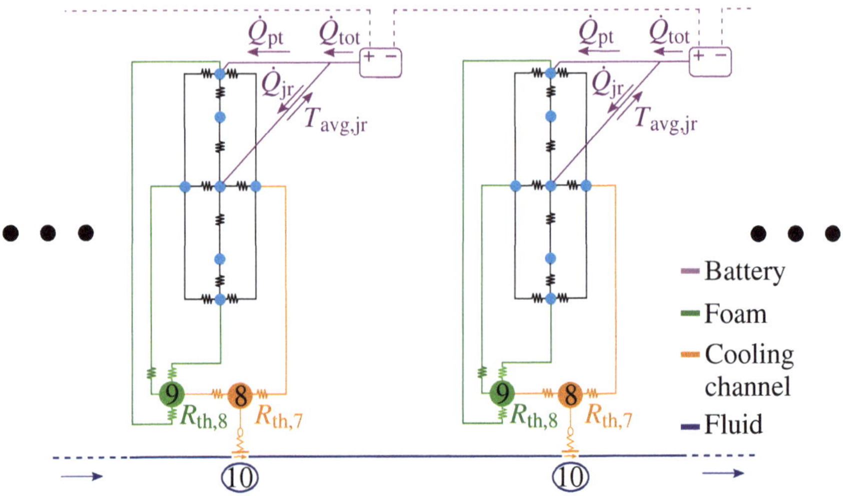

Figure 4.10: Simulation framework for the 1D module section model, exemplified by two LIB cells.

foam, heat conduction within the foam is considered negligible in this study. Consequently, the corresponding thermal resistance is set to $R_{th,8} = 10^5$ K/W.

The coolant inlet boundary conditions are defined by the mass flow rate, temperature and pressure. To model the heat transfer between the coolant and the cooling channel using the thermal-hydraulic internal flow module with heat exchange, Nusselt number correlations, pressure drop correlations and specifications of the fluid channel geometry are required. The geometric data are derived from the 3D design, while the correlations are obtained from the detailed 3D CFD heat sink simulations.

4.4 Full module level modeling

The 24-cell module section can be upscaled to represent the full module level. As illustrated in Figure 4.11, it is not necessary to simulate every cell in the full pack. Instead, a model reduction strategy is employed to minimize computational effort during life-cycle simulations. For this purpose, the symmetry of the defined reference battery pack and BTMS is utilized to reduce the investigation to just one module. Additionally, each module consists of seven wavy mini-channel heat sinks, each cooling two linear arranged cylindrical LIB cell rows. Assuming that the thermal contribution from the battery casing is negligible, it is sufficient to analyze only one wavy mini-channel heat sink along with its associated cells to effectively evaluate the BTMS concept's temperature-related aging inhomogeneities. As a result of this model reduction, the coolant volume flow rate for a single stacked wavy mini-channel heat sink must be derived from the total BTMS volume flow rate.

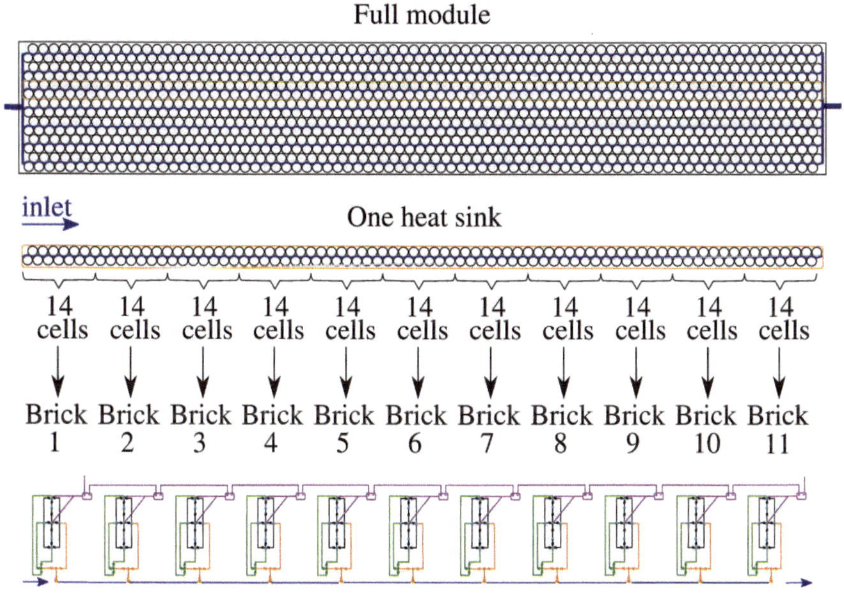

Figure 4.11: Schematic of the applied model reduction on the full module level.

To further reduce computational effort, multiple cells are aggregated into a single electro-thermal representation, hereafter referred to as a 'brick'. In this approach, both the fluid channel geometry and the heat exchange surface parameters must be scaled to the heat sink section attributed to the number of cells represented by each brick. In the final model, each brick comprises 14 LIB cells. Consequently, the model consists of 11 bricks, corresponding to a total of 154 LIB cells cooled by a single stacked wavy mini-channel heat sink. This configuration represents a battery module with an approximate length of ≈ 1.74 m. Due to the aggregation into bricks, the predicted LIB temperature for each brick represents the average mean jelly roll temperature of the 14 constituent cells.

In this thesis, all 3D simulations are conducted using Simcenter STAR-CCM+, which employs the finite volume method to solve the governing equations for energy, mass and momentum conservation. In the transient 3D simulations, a time step of $\Delta t = 1$ s is employed, with five inner iterations per time step demonstrating sufficient numerical convergence. The corresponding 1D simulations are conducted with Simcenter AMESIM, utilizing the software's built-in solver. For further details, the reader is referred to the respective user manuals [97, 98].

5 Validation

This chapter presents the validation of the modeling approaches used in this thesis, including the ECM-coupled 3D and 1D thermal models, the Arrhenius based aging model and the heat transfer characteristics of the wavy mini-channel flow CFD simulations.

For this purpose, several key validation metrics are introduced. The mean error (ME) is defined as

$$\mathrm{ME} = \frac{\sum_{i=1}^{n} \left(x_{i,\mathrm{exp}} - x_{i,\mathrm{sim}} \right)}{n} \tag{5.1}$$

for n comparison steps between experimental and simulation results. Furthermore, the root mean squared error (RMSE) is given as

$$\mathrm{RMSE} = \sqrt{\frac{\sum_{i=1}^{n} \left(x_{i,\mathrm{exp}} - x_{i,\mathrm{sim}} \right)^2}{n}} \tag{5.2}$$

and the root mean squared relative error (RMSRE) as

$$\mathrm{RMSRE} = \sqrt{\frac{1}{n} \sum_{i=1}^{n} \left(\frac{x_{i,\mathrm{exp}} - x_{i,\mathrm{sim}}}{x_{i,\mathrm{exp}}} \right)^2} \tag{5.3}$$

both presenting common evaluation metrics to discuss the accuracy of simulative modeling approaches.

5.1 Cell level validation

To validate the ECM-coupled 3D and 1D thermal models, the experimental measurement setup shown in Figure 5.1 is employed. In this setup, the LIB cell is mounted in the cylindrical Feinmetall cell holder to provide electrical connection and measurement capability. To avoid introducing additional contact surfaces, only the positive and negative terminals are in contact with the cell holder. A 10 mm air gap at the bottom of the cell holder ensures no additional heat conduction occurring between the cell holder and the LIB cell.

M. Auch, *Holistic Modeling Framework for the Investigation of Innovative Battery Thermal Management Concepts*, Wissenschaftliche Reihe Fahrzeugtechnik Universität Stuttgart, https://doi.org/10.1007/978-3-658-52612-2_5

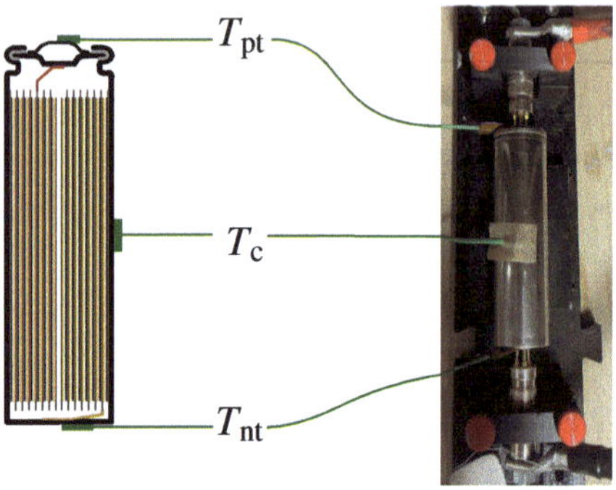

Figure 5.1: Experimental measurement setup for the electro-thermal characterization. Adapted from [79].

For thermal measurements, self-adhering type K thermocouples with an oil-bath calibrated accuracy of 0.1 K are used. The thermal behavior of the LIB cell is recorded at the cell mantle T_c, positive terminal T_{pt} and negative terminal T_{nt}. This setup presents a good balance between accurately capturing the temperature distribution around the cell surface and minimizing the number of thermocouples attached to the small outer surface of the LIB cell, which could otherwise influence the heat dissipation behavior in the laboratory setup.

5.1.1 Load profiles

Ensuring a representative range of BEV operating conditions, four different load profiles are defined, which include a passive warm-up, a highly dynamic World Harmonized Light Duty Vehicles Test Procedure (WLTC), a fast charge and a slow charge case. While the warm-up, fast charge and slow charge scenarios can be defined directly at the cell level, the WLTC profile requires additional assumptions based on vehicle characteristics.

At low temperatures, charging and discharging rates are limited due to accelerated aging effects, primarily lithium plating, which necessitate pre-heating of LIB cells. Due to the laboratory environment, a passive warm-up scenario is used for validation instead of active pre-heating. In this scenario, the cell is passively heated from 5 °C to a minimum of 15 °C. To ensure thermal and chemical equilibrium at the initial temperature, the cell is rested for 3 hours. After equilibrium is reached, the climate chamber is turned off and its door opened, exposing the LIB cell to the current ambient room temperature ($\approx 25\,°C$).

The three electro-thermal load cycles follow a similar experimental procedure. For both fast and slow charging, the LIB cell is initially set to an SOC of 20 % and rested at an equilibrium temperature of 25 °C. The climate chamber is then turned off to avoid any influence from its ventilation system on temperature measurements. A constant-current constant-voltage (CC-CV) profile is subsequently applied. In the fast charge scenario, a constant current of 2 C is used, while a current of 0.33 C is chosen for slow charging. The constant-current phase is terminated when the cell reaches a maximum voltage of $U = 4.15\,V$, after which it switches to the constant-voltage phase. This voltage limit is set slightly below the maximum specified in the datasheet (4.2 V), as commonly done by BEV manufacturers to mitigate accelerated degradation in the high SOC range [99].

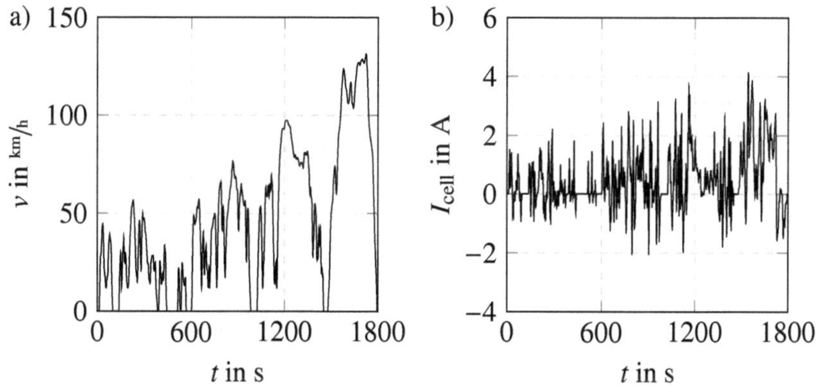

Figure 5.2: WLTC class 3 test profile: a) Velocity over time on vehicle level and b) Current over time on cell level.

The highly dynamic load profile is derived from the WLTC Class 3 profile shown in Figure 5.2 a) [100]. Using the reference vehicle assumptions summarized in Appendix Table E.1 and applying Equations (2.1), (2.2) and

(2.3), the cell current over time for the WLTC profile is calculated and shown in Figure 5.2 b). For this scenario, the LIB cell is set to an initial SOC of 25 % and an equilibrium temperature of $T = 25\,°\mathrm{C}$. Once equilibrium is reached, the climate chamber is turned off, and the WLTC profile is initiated.

5.1.2 3D thermal model

The final step in validating the ECM-coupled 3D thermal cell model consists of defining the boundary conditions in the laboratory setup. These are established as a superposition of heat convection and radiation [66]. During the experimental measurements, the climate chamber is turned off prior to initiating the corresponding load profile. As a result, convective heat transfer is assumed to occur via natural convection.

The boundary conditions are further differentiated between the cell mantle, positive terminal and negative terminal to account for potential variations in their respective surface temperatures. Of particular importance is the positive terminal boundary condition. In the simplified model depicted in Figure 4.4 c), the outer surface of the positive terminal is represented as a flat surface. However, in the actual LIB cell, a gap exists between the upper thin plate and the remainder of the cap, as shown in Figure 4.4 a). To account for the additional surface area introduced by this cap design, the boundary condition normal n to the positive terminal surface is increased by a factor of two, resulting in

$$\mathrm{BC}_{\mathrm{pt}} : -k\frac{\partial T}{\partial n} = 2\,(h_{\mathrm{conv}} + h_{\mathrm{rad}})\,(T_{\mathrm{s,avg,pt}} - T_{\mathrm{amb}}) \tag{5.4}$$

where $T_{\mathrm{s,avg,pt}}$ is the surface-averaged temperature of the positive terminal. The boundary conditions normal to the cell mantle BC_{c} and negative terminal $\mathrm{BC}_{\mathrm{nt}}$ surfaces are given by

$$\mathrm{BC}_{\mathrm{c}} : -k\frac{\partial T}{\partial n} = (h_{\mathrm{conv}} + h_{\mathrm{rad}})\,(T_{\mathrm{s,avg,c}} - T_{\mathrm{amb}}) \tag{5.5}$$

and

$$\mathrm{BC}_{\mathrm{nt}} : -k\frac{\partial T}{\partial n} = (h_{\mathrm{conv}} + h_{\mathrm{rad}})\,(T_{\mathrm{s,avg,nt}} - T_{\mathrm{amb}}) \tag{5.6}$$

where $T_{s,avg,c}$ and $T_{s,avg,nt}$ represent the surface-averaged temperatures of the cell mantle and negative terminal, respectively.

The convective heat transfer coefficients h_{conv} for natural convection are calculated for the cell mantle using Equation (2.24) and for the terminals using Equation (2.25). The simplified radiative heat transfer coefficient is calculated using Equation (2.17), assuming an emissivity value of $\epsilon = 0.45$ for both the LIB cell casing and the internal climate chamber surface material, which lies within the literature range for steel surfaces [101]. The ambient temperature T_{amb} is measured during each experimental profile and incorporated directly into the simulation to define the time-dependent boundary conditions. These parameters are updated and solved at each simulation time step $\Delta t = 1$ s.

The previously introduced boundary conditions assume ideal natural convective and radiative heat transfer around a surface exposed to the ambient environment. However, in the actual electro-thermal experimental setup, the LIB cell is placed inside a cylindrical cell holder. Consequently, the passive warm-up scenario is investigated first to evaluate the potential influence of the LIB cell's position within the holder on the boundary conditions.

Passive warm-up

For this purpose, there are two different warm-up scenarios investigated. In the first setup, case 1, the LIB is located in the center of the climate chamber (ideal natural convective and radiative heat transfer) and in the second setup, case 2, the LIB is positioned inside the cylindrical cell holder (actual heat transfer conditions, as illustrated in Figure 5.1). The results for both cases are illustrated in Figure 5.3.

The scenario of the LIB hanging in the center of the climate chamber (case 1) is discussed first. As shown in Figure 5.3 a), the temperature increases similarly at all three measurement points. It can be observed, that the temperature increases the fastest at the positive terminal, then the negative terminal and lastly, the mantle. This is elucidated by analyzing the heat conduction and convection phenomena within the LIB cell. The temperature measurement location at the positive terminal is subject to increased heat convection and a low mass. Furthermore, the electrical insulation limits heat

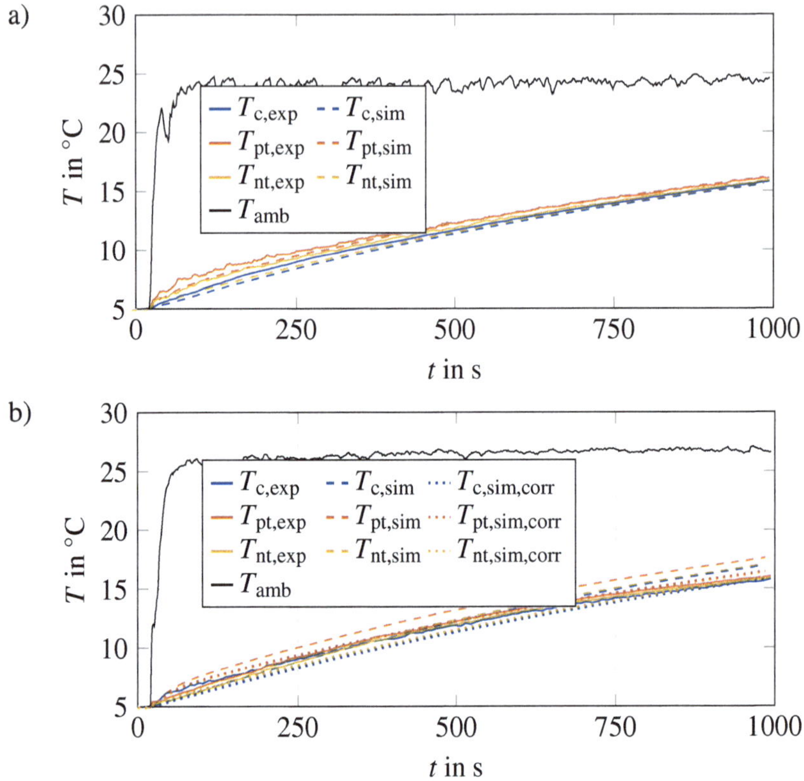

Figure 5.3: Experimental and simulated temperature T at three measurement points: a) Case 1 warm-up and b) Case 2 warm-up.

conduction to the jelly roll, resulting in the observed fastest temperature increase at the positive terminal. The difference between negative terminal and mantle can also be explained by the electrical insulation layer, which is located between negative terminal and jelly roll.

Figure 5.3 a) and Table 5.1 are used to discuss the simulation results. With RMSE < 0.5 K at all three measurement positions the simulation and experimental results show a very good agreement. However, a difference between the measurement and simulation can be observed in the first 150 s.

The reason can be identified by considering the test procedure. After the LIB has been rested, the climate chamber is turned off and the climate chamber door is immediately opened. This door opening procedure can cause a temporary fluid movement resulting in forced convection and therefore, increased heat transfer around the LIB cell in the early time frame. As the climate chamber door remains closed in all subsequent electro-thermal experimental cases, this effect is considered negligible.

Table 5.1: MEs and RMSEs (RMSREs) of simulation and measurement results for the warm-up cases of the MPH-TR model. RMSREs are always referenced to the time-dependent experimental value in °C.

Parameter	Warm-up Case 1	Warm-up Case 2 w/o correction	Warm-up Case 2 with correction
ME_c in K	0.32	-0.29	0.46
ME_{pt} in K	0.14	-1.02	-0.17
ME_{nt} in K	0.41	-0.51	0.27
$RMSE_c$ in K	0.35 (4.00 %)	0.64 (5.43 %)	0.54 (6.31 %)
$RMSE_{pt}$ in K	0.27 (3.29 %)	1.06 (9.45 %)	0.25 (3.18 %)
$RMSE_{nt}$ in K	0.46 (5.24 %)	0.60 (4.66 %)	0.35 (3.13 %)

In Figure 5.1 b), the results of the case 2 warm-up scenario are shown, where the placement of the LIB has been changed to inside the cell holder. In this case, the LIB doesn't exhibit ideal natural convection and radiation boundary conditions because its lower side is faced towards the cell holder's top surface. As a result, the heat transfer is overestimated by the applied correlations. To consider this influence, a correction of the natural convection and radiation boundary conditions is proposed. The correction factor BC_{corr} is applied to all boundary conditions as

$$BC_{corr,i} = BC_{corr}\, BC_i \tag{5.7}$$

with an empirical determined correction factor of $BC_{corr} = 0.85$. Hence, the changed positioning of the LIB cell from the center of the climate chamber to the cell holder decreases the heat transfer to the environment by 15 %. The results for the simulation with and without the empirically determined correction factor are shown in Table 5.2. The empirically corrected boundary conditions are used for all electro-thermal simulations.

Electro-thermal load profiles

With the boundary conditions established, the thermal model can be validated for the three electro-thermal load profiles. Figure 5.4 a) and b) show the experimental results for the fast charge scenario, which are discussed first. The maximum temperature is observed shortly after the constant current phase at the positive terminal measurement point, with $T_{\mathrm{exp,max,pt}}(t = 623\,\mathrm{s}) = 63.9\,°\mathrm{C}$. During the constant current phase, the temperature increase at the positive terminal is significantly higher compared to the cell mantle and negative terminal measurement points, which is mainly attributed to the positive tab ohmic heat generation. The overall temperature rise at the cell mantle and negative terminal positions is similar, although the temperatures at the negative terminal are slightly lower than those at the mantle. At all three measurement positions, the temperature decreases shortly after the constant current phase ends.

As shown in Table 5.2, the simulation results at the cell mantle and negative terminal exhibit very good agreement with the experimental measurements, with RMSEs of $\mathrm{RMSE_c} = 0.45\,\mathrm{K}$ and $\mathrm{RMSE_{nt}} = 0.75\,\mathrm{K}$, respectively. Although the predicted temperature at the positive terminal also shows good overall agreement, a higher RMSE of $\mathrm{RMSE_{pt}} = 2.47\,\mathrm{K}$ is observed. At this measurement position, the simulation underestimates the temperature during the constant current phase and overestimates it during the constant voltage phase.

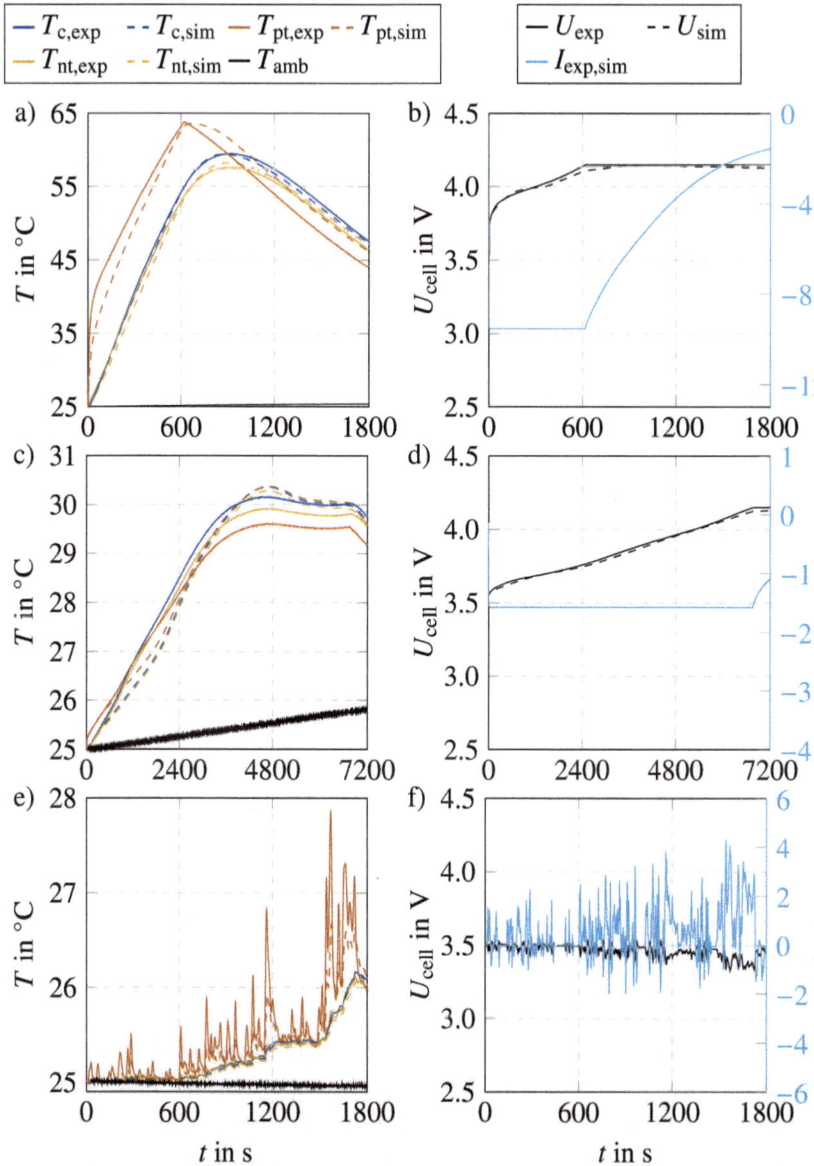

Figure 5.4: Experimental and simulated temperature and voltage for fast charge (a),b)), slow charge (c),d)) and WLTC at SOC=25 % (e),f)).

This behavior can be attributed to the simplified 3D thermal modeling approach. In the experimental setup (Figure 5.1), the temperature is measured at the thin upper steel plate of the positive terminal. Due to its low mass, this upper steel plate responds more rapidly to heat generation and convection compared to the simplified cap region in the developed thermal model. Nevertheless, the thermal behavior of the LIB cell is still considered to be sufficiently modeled, especially considering the reduced computational effort. The higher temperature increase at the positive terminal further results in the maximum temperature gradient inside the jelly roll (active material) occurring along the cell length under laboratory heat transfer conditions, as illustrated in Appendix Figure E.1. The voltage behavior is modeled accurately by the developed ECM, with a $RMSE_U$ of 19 mV.

Table 5.2: MEs and RMSEs (RMSREs) of simulation and measurement results for the electro-thermal cases of the MPH-TR model.

Parameter	Fast charge	Slow charge	WLTC at SOC=25 %
ME_c in K	0.36	0.17	0.02
ME_{pt} in K	-0.42	-0.27	0.12
ME_{nt} in K	0.5	0.05	0.03
$RMSE_c$ in K	0.45 (0.90 %)	0.33 (1.17 %)	0.02 (0.09 %)
$RMSE_{pt}$ in K	2.47 (5.13 %)	0.46 (1.56 %)	0.23 (0.84 %)
$RMSE_{nt}$ in K	0.75 (1.78 %)	0.31 (1.09 %)	0.04 (0.16 %)
ME_U in mV	14	13	-4
$RMSE_U$ in mV	19 (0.47 %)	14 (0.36 %)	4 (0.12 %)

In Figure 5.4 c) and d), the experimental measurement results for the slow charge scenario are presented. The overall temperature increase is lower compared to the fast charge scenario, primarily due to the reduced heat generation resulting from the lower applied current. The RMSEs for the slow charge scenario are all below 0.5 K at the three measurement positions, indicating very good agreement with the experimental results.

However, between 1000 and 3000 s, a slight deviation between the measurements and the simulation can be observed. This deviation is attributed to the entropic coefficient, which has an increased influence on the total generated heat at lower compared to higher C-rates [102]. Because the determination of the entropic coefficient requires a high voltage measurement resolution, measurement uncertainties of the used test bench can affect the results. This is assumed to be the dominant factor responsible for the observed deviation. The voltage is modeled with an $RMSE_U$ of 14 mV, indicating accurate prediction of the voltage behavior.

The WLTC results are shown in Figure 5.4 e) and f). A highly sensitive temperature response to the applied current pulses can be observed at the positive terminal. Minor fluctuations are also present at the negative terminal due to the heat generation in the negative tab, but these are significantly smaller compared to the positive terminal, further supporting the MPH-TR approach of attributing negative tab heat generation to the jelly roll region. Overall, the WLTC scenario is not thermally challenging, as the maximum temperature increase at the cell mantle remains below 1.5 K and the maximum temperature increase at the positive terminal is below 3 K. The model's accuracy compared to the experimental measurements is very good at all three measurement positions, with RMSEs below 0.3 K, as shown in Table 5.2. Furthermore, the developed electro-thermal model accurately captures the highly dynamic voltage behavior of the WLTC scenario ($RMSE_U$ of 4 mV).

5.1.3 1D thermal model

In addition to using the experimental measurement data, the 1D thermal resistance network model (as previously illustrated in Figure 4.6) is validated by comparing the predicted temperature of the mean jelly roll node (1) with the volume-averaged jelly roll temperature $T_{avg,jr}$ computed from the 3D model. Since this node is the input for the electrical model, this approach ensures accurate prediction of the average jelly roll temperature while simplifying the thermal modeling approach suitable for life-cycle studies on the module level. The previously introduced radiative and convective heat transfer boundary conditions are applied to the positive terminal node (4), the two casing mantle nodes (6,7) and the negative terminal node (5). Since the overall thermal behavior is similar to the 3D simulation results, they are only illustrated for the fast charge scenario in Figure 5.5.

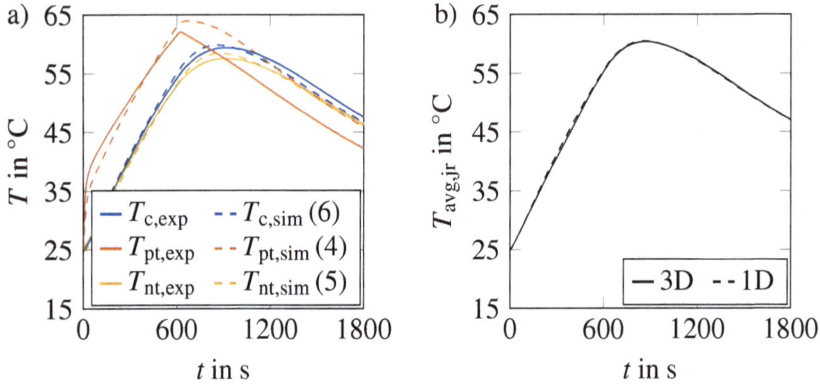

Figure 5.5: Temperature in the defined fast charge scenario: a) Experimental and 1D simulation results and b) Simulated 3D volume-averaged and 1D node (1) temperature results.

As presented in Table 5.3, the 1D model demonstrates good accuracy compared to the experimental data, with RMSEs constantly below 1 K for all load cycles at all measurement positions except for the positive terminal in the fast charge scenario. Here, the same reasoning as for the 3D model applies for the observed discrepancies. The additional comparison of the node (1) to the volume-averaged jelly roll temperature of the 3D model demonstrates very good accuracy with RMSEs below 0.3 K. Consequently, the 1D model is considered validated not only for the experimental measured surface temperatures but also for the prediction of the mean jelly roll temperature.

5.1.4 Cell-to-cell variation

The previous validation has been conducted for one single sample of the studied LIB cell. To ensure the approach of up-scaling single cell validated models to the module level, four additional cells have been tested for the

Table 5.3: MEs and RMSEs (RMSREs) of simulation and measurement results for all studied scenarios of the ECM-coupled 1D thermal cell model.

Parameter	Warm-up Case 2 with correction	Fast charge	Slow charge	WLTC at SOC=25 %
ME_c in K	0.50	0.10	0.39	-0.02
ME_{pt} in K	0.32	-2.30	-0.08	0.08
ME_{nt} in K	0.23	0.10	0.32	0.01
$RMSE_c$ in K	0.59 (6.86 %)	0.77 (1.48 %)	0.45 (1.56 %)	0.03 (0.10 %)
$RMSE_{pt}$ in K	0.38 (3.56 %)	3.44 (6.93 %)	0.29 (1.00 %)	0.22 (0.82 %)
$RMSE_{nt}$ in K	0.30 (2.73 %)	0.59 (1.16 %)	0.39 (1.37 %)	0.04 (0.14 %)
RMSE $T_{avg,jr}$ to 3D in K	0.02 (0.13 %)	0.27 (0.42 %)	0.15 (0.43 %)	0.04 (0.11 %)
ME_U in mV	-	12	43	3
$RMSE_U$ in mV	-	17 (0.41 %)	45 (1.18 %)	6 (0.16 %)

fast charge and WLTC scenario to evaluate cell-to-cell variations of the thermal behavior. Here, the assumption has been made, that measuring at one position (T_c) is enough to show the similarity of the thermal behavior. The temperature measurement results for the fast charge and WLTC scenario of all five cells are shown in Figure 5.6. The previously discussed and validated experimental data is included as Cell 1.

Analyzing the experimental measurement results, both the thermal and voltage behaviors of the five studied cells exhibit similar trends. The voltage response shows very good agreement, with average absolute deviations of less than 3 mV for all five cells during the fast charge scenario when compared to the arithmetic mean. For the temperature behavior of the fast charge scenario, the highest average absolute deviation occurs in Cell 1, with

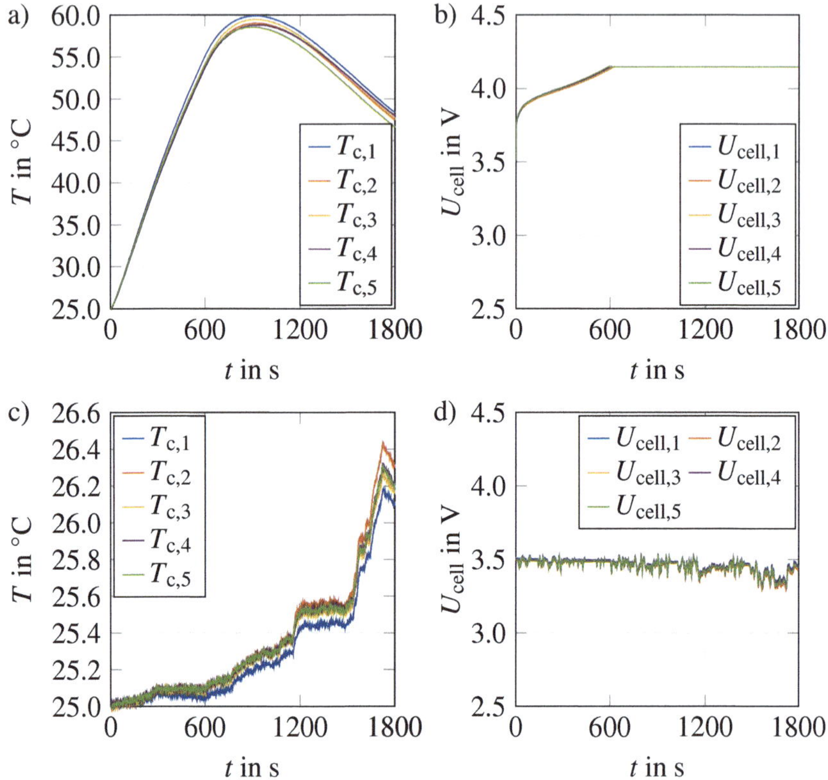

Figure 5.6: Measured cell mantle position temperature T_c and cell voltage of five cells for fast charge (a),b)) and WLTC at SOC=25 % (c),d)). Index 1 represents the previously discussed and compared cell results.

an RMSE of 0.67 K. In contrast, for the WLTC scenario, the deviation for Cell 1 is even lower, with an RMSE of just 0.05 K. The previously presented simulation deviations at the cell mantle, which have RMSEs below 0.80 K, are in a similar range. Consequently, the simulation accuracy is comparable to the cell-to-cell variation observed in the experimental measurements. As a result, the upscaling from single-cell validated models to module-level simulations is assumed to be valid.

With the successful electro-thermal validation across multiple load cycles and measurement positions, the cell-level validation is completed by deriving the aging model parameters.

5.1.5 Aging model

For the parameterization of the Arrhenius-based aging model presented in Section 4.2.4, an experimental aging study is conducted to characterize the aging behavior of the LIB cells under controlled cycling conditions. The Basytec XCTS Mk 11 test bench is used for this purpose, enabling the simultaneous testing of five cells.

In aging studies, key boundary conditions include ambient temperature, applied current rate, selected SOC levels and charge/discharge cut-off voltages [103]. Since full cycle aging studies until the EOL require significant time, these experiments inherently include calendar aging effects. However, separating calendar and cyclic aging effects would require an independent calendar aging study. Since both aging effects occur simultaneously in real-world applications, this thesis does not attempt to separate them, assuming that calendar aging is sufficiently included in the experimental study [104].

The aging tests are conducted in a climate chamber set to $T_{\mathrm{amb}} = 25\,°\mathrm{C}$ to ensure stable ambient conditions. The charge/discharge cycle consists of a CC-CV charging phase at 0.5 C, followed by a constant-current discharge at 0.5 C. To reflect typical BEV operation, the cells are cycled within a restricted voltage range, rather than the full nominal range [99]. The upper voltage limit for the CV phase is set to $U_{\mathrm{max,aging}} = 4.15\,\mathrm{V}$ and the lower voltage limit is set to $U_{\mathrm{min,aging}} = 3.0\,\mathrm{V}$, corresponding to an SOC range of 4-97 % at BOL for the studied LIB cell.

Aging is monitored by tracking capacity fade and internal resistance growth through periodic capacity and HPPC tests. These tests are conducted initially and after every 100 load cycles. The capacity test employs a 0.2 C discharge current, while the HPPC test follows the procedure defined in [29]. Aging progression is quantified based on cumulative charge throughput, calculated as the total ampere-hours passed through each cell at each aging checkpoint. The resulting capacity loss data points for the five tested cells are shown in Figure 5.7, labeled as Cell A1 to A5.

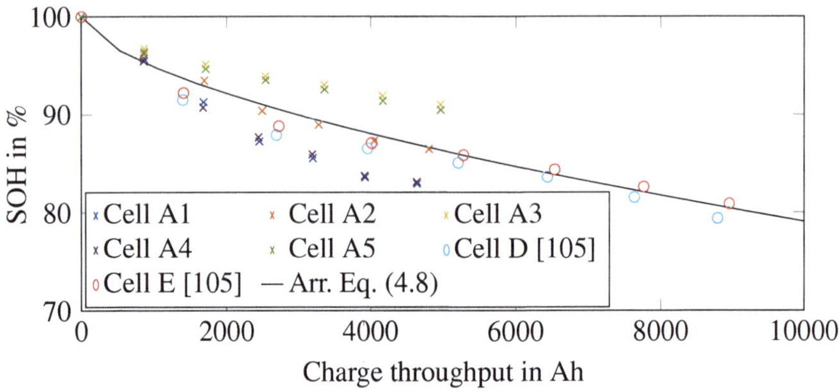

Figure 5.7: SOH over charge throughput for the in-house aging study (Cells A1-A5), Kirkaldy [105] Cell D/E and the Equation (4.8) based Arrhenius-fit.

Analyzing the capacity loss data points, between the five tested cells a significant variation can be observed (up to $\Delta SOH = 8\,\%$ at the final aging checkpoint). This variation is present although the same cycling profile is applied and the average measured surface temperature of all five cells being within 1 K of the set ambient temperature $T = 25\,°C$, which is due the moderate C-rate chosen for the aging study.

In general, the variations are not surprising, as other research also reports a significant variation from cell to cell and recommends testing multiple cells at the same conditions to derive accurate average capacity loss values [106, 107]. These variations are attributed to the random aging effect and the studied LIB cells being not sourced from the original manufacturer but from a supplier. Consequently, the tested cells may not fulfill the BEV manufacturing tolerances and/or there is no information about potential storage conditions prior to delivery (storage time, temperature, SOC), which could cause different calendar aging behaviors of the studied cells.

As a consequence, in addition to applying a parametric curve fit through the measured experimental data points to derive the capacity loss model parameters presented in Equation (4.8), two aging data sets of the literature [105] for the same cell type, referenced as Cell D and E, are used for validation of the derived model (0.3 C CC-CV charging, 1 C discharging, T = 25 °C). As shown in Figure 5.7, the arithmetic mean capacity loss behavior derived from the own aging study is in good agreement with the literature data. However, in actual BEVs, especially during driving (discharging) there will be no constant loads applied to the LIB cell. Consequently, the true battery aging in BEVs can vary from laboratory tests, as constant power cycles cause accelerated aging compared to dynamic loads [105, 108].

The chosen Arrhenius-based aging model presents an integrated dependency of the capacity loss behavior on the operating temperature. In this thesis, the parametric curve fit of Equation (4.8) is conducted to consider the SEI-layer growth as dominant aging effect for temperatures of 25 °C and above [6]. Further validation data sets at both higher and lower temperatures would enhance the model's accuracy and enable also including an aging model for the lithium-plating effect. However, due to the required duration of these additional tests, this has not been further pursued. Instead, it is assumed that the Arrhenius-based approach sufficiently captures the temperature dependency.

Furthermore, the consideration of the C-rate dependent aging is limited due to the availability of only one C-rate data set. As a result, the C-rate effect is integrated following Wang et al. [92], by modifying the activation energy with a C-rate-dependent factor. Generally, with an increasing C-rate the activation energy of the aging model decreases [92], resulting in accelerated aging. According to Iurilli [109], the studied LIB cell exhibits superlinear degradation at typical fast charging currents of 2 C. Therefore, 2 % reduction in the aging model activation energy resulting from an increase of the C-rate from 0.5 C to 2 C is assumed for this study. However, as also the crossover temperature for the dominant aging effect may shift with changing C-rates, this assumption must be interpreted with caution [42, 93]. The derived parameters for the capacity loss model are given in Table 5.4.

The additional conducted HPPC test results during the aging study are used to determine the resistance increase during each aging checkpoint. By using Equation (4.9), the resistance increase can be empirically correlated to the capacity loss behavior. The resistance increase is split into ohmic

Table 5.4: Derived capacity loss aging model parameters.

Parameter	B_{cyc}	$E_{a,cyc}$ in J/mol	z_{cyc}	α
Value	14337.5	30278.1	0.6084	83.5488

resistance R_0 and polarization resistance $R_{p,1,2}$, dependent on the derived ECM resistance values. For this purpose, the average resistance increase over all SOC measurement points is chosen, with the results being shown in Figure 5.8.

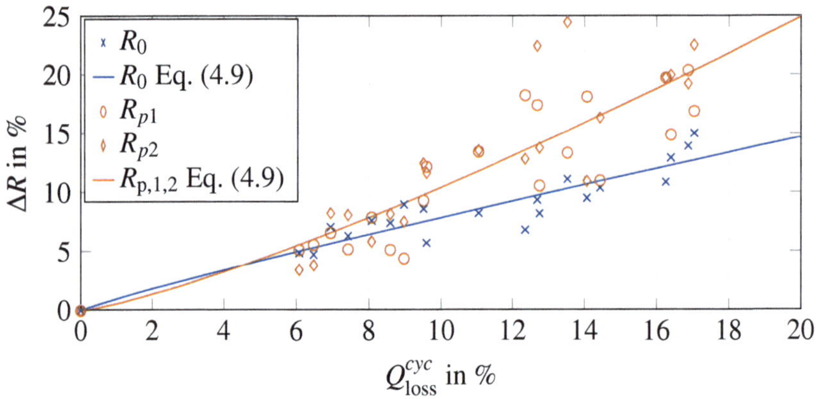

Figure 5.8: Resistance increase ΔR over capacity loss Q_{loss}^{cyc} of the conducted aging study results (Cells A1-A5) and the Equation (4.9) based fit split in ohmic R_0 and polarization resistance $R_{p1,2}$.

Similar to the capacity loss data, a noticeable cell-to-cell variation is observed in the resistance increase measurements. This variation is more pronounced for the polarization resistance $R_{p,1,2}$, whereas the variation in the ohmic resistance R_0 is relatively small. The increase in polarization resistance $R_{p,1,2}$ as a function of capacity loss follows a power-law dependence, while the dependence of the ohmic resistance R_0 is closer to linear. Consistent with the capacity loss aging model, a parametric curve fit is applied to all measurement data, distinguishing between the two components. The parameters of the resulting resistance increase aging model are listed in Table 5.5.

Table 5.5: Derived resistance increase aging model parameters.

Parameter	$\alpha^R_{\mathrm{cyc}}(R_0)$	$\alpha^R_{\mathrm{cyc}}(R_{\mathrm{p1,p2}})$	$z^R_{\mathrm{cyc}}(R_0)$	$z^R_{\mathrm{cyc}}(R_{\mathrm{p1,p2}})$
Value	0.9844	0.5815	0.9041	1.2550

Finally, entropic coefficient measurements are conducted at BOL and at each aging checkpoint. As illustrated in Figure 5.9 a), the SOC-dependent entropic coefficient shows slight variations with increasing cycle count. In particular, between BOL and cycle 300, the entropic coefficient appears shifted toward lower SOCs. However, this effect is attributed to changes in the OCV-SOC relationship. As the LIB cells age, the OCV corresponding to a given SOC decreases, resulting in lower OCV values at the same SOC.

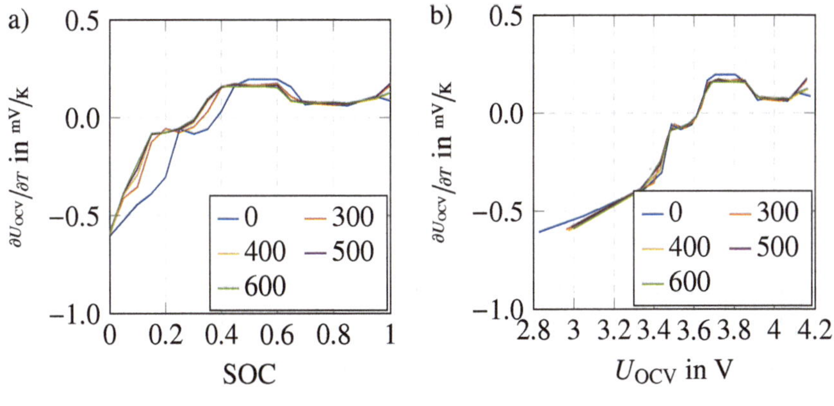

Figure 5.9: Entropic coefficient $\partial U_{\mathrm{OCV}}/\partial T$ of aging study load cycle checkpoints over: a) SOC and b) U_{OCV}.

In contrast, as shown in Figure 5.9 b), when evaluating the entropic coefficient as a function of OCV, no observable change is detected over the full aging study. Consequently, when analyzing LIB aging with a focus on temperature and heat generation, the entropic coefficient should be modeled as a function of OCV rather than SOC. Although Tendera et al. [110] indicate that LIB aging can also impact the thermal properties, this effect is neglected in the investigation of this thesis.

With the developed aging model, all cell-level models have been successfully validated and the upscaling from single-cell models to the module level has been confirmed by demonstrating similar thermal behavior across multiple cells subjected to identical load profiles. As a result, this thesis does not further address electro-thermal validation at the module section or full module level. Instead, particular emphasis is placed on accurately modeling the heat transfer characteristics of the heat sink within the 3D and 1D model.

5.2 Module section level validation

At the module section level, the validation of the heat transfer characteristics is conducted on the isolated heat sink, excluding the remaining module components. This approach enables a focused validation of the heat sink's thermal performance against reference data from the literature. For this purpose, key heat transfer characteristics, in particular the local Nusselt number Nu_{lo} the average Nusselt number Nu_{avg} and the pressure drop Δp, are used to investigate the coolant flow behavior in straight and wavy mini-channel heat sinks, as introduced in Section 2.4.

5.2.1 Heat transfer characteristics

Due to the lack of available correlations for wavy mini-channels with rectangular cross-sections in the literature, the single straight mini-channel presented in Section 4.3.1 is chosen for the validation of the CFD-predicted heat transfer characteristics. The correlations proposed by Lee and Garimella [68] are used as reference data for this validation.

The flow is assumed to be steady-state, incompressible, laminar, with constant fluid properties and thermally developing conditions. For the thermal boundary conditions on the cooling channel outer surface, the H1 thermal boundary condition is applied, which assumes a constant wall heat flux in the axial direction and a constant wall temperature in the peripheral direction. This boundary condition represents a common simulation approach for evaluating isolated heat sink performance in cooling channels with high thermal conductivity materials, such as aluminum in the reference BTMS heat sink design [111]. The validation results for the local Nusselt numbers Nu_{lo} at various dimensionless channel length positions y^* are summarized in Table 5.6.

Table 5.6: Validation of straight rectangular mini-channel flow with H1 BC and EGL 50:50 at Re = 187, T_{in} = 25 °C and q = 8000 W/m².

	This study	Lee and Garimella [68]	Deviation in %
Nu_{lo} at y^* = 0.005	6.78	6.77	0.08
Nu_{lo} at y^* = 0.01	5.51	5.54	-0.59
Nu_{lo} at y^* = 0.1	3.67	3.64	0.78
	This study	Analytical Solution [112]	Deviation in %
Friction factor f	0.0764	0.0766	0.26

The CFD simulation results show good agreement with the correlations provided by Lee and Garimella [68], exhibiting a maximum absolute deviation of 0.78 % and a mean absolute deviation of 0.48 %. To validate not only the heat transfer characteristics but also the pressure drop prediction, the CFD-simulated pressure loss is compared to the analytical solution of Bennett [112] using the Darcy friction factor. As presented in Table 5.6, with a deviation of 0.26 % the Darcy friction factor of the CFD simulation agrees well with the literature value.

Although there are no existing correlations for wavy channels with rectangular cross-sections, the CFD-predicted results are compared to the correlation for the average Nusselt number Nu_{avg} provided by Khoshvaght-Aliabadi et al. [113] for wavy channels with square cross-sections. This correlation is defined as

$$Nu_{\text{avg}} = 0.1105\, Re^{0.6909}\, Pr^{0.4743} \left(\frac{A}{d_\text{h}}\right)^{0.0027} \left(\frac{\lambda}{d_\text{h}}\right)^{-0.2714} \tag{5.8}$$

where A denotes the amplitude and λ the wavelength of the wavy channel geometry. The comparison results are summarized in Table 5.7.

Table 5.7: Validation of wavy rectangular mini-channel flow with H1 BC and EGL 50:50 at $T_{\text{in}} = 25\,°\text{C}$ and $q = 8000\,\text{W}/\text{m}^2$.

Reynolds Number Re	Prandtl Number Pr	Nu_{avg} this study	Nu_{avg} [113]	Deviation in %
50	25.94	4.30	3.91	9.97
150	25.94	8.90	8.35	6.59
200	25.94	11.87	10.18	16.60
250	25.94	14.46	11.88	21.72
350	25.94	17.28	14.99	15.28

For the comparison, a total of five different Reynolds numbers $Re = 50, 150, 200, 250, 350$ are investigated. The simulation results consistently predict higher average Nusselt numbers, with a mean absolute deviation of 14.03 %. This deviation is expected, given the aspect ratio of the rectangular cross-section used in this study, in contrast to the square geometry considered in the correlation. Considering that the correlation proposed by Khoshvaght-Aliabadi et al. [113] itself exhibits a mean absolute deviation of 7.3 % relative to their experimental data, the CFD-predicted Nusselt numbers for the wavy channel with a rectangular cross-section appear reasonable.

In the actual reference BTMS design, 30 wavy mini-channels are stacked, as illustrated in Figure 4.9. Consequently, the influence of stacking multiple channels, as opposed to using a single wavy mini-channel, on their heat transfer characteristics is investigated ($Re = 187, T_{\text{in}} = 25\,°\text{C}$, EGL 50:50). The comparison reveals that for the average Nusselt number the deviation is 0.43 % and for the local Nusselt number, the deviation is 0.38 %. These findings indicate that stacking multiple channels does not significantly influence the heat transfer characteristics. This observation is attributed to the high thermal conductivity of the aluminum cooling channel fins, which ensures that the thermal boundary conditions around each cooling channel fluid in the stacked configuration remain comparable to those in the single-channel configuration.

5.2.2 Heat sink correlations

As outlined in the holistic thermal modeling approach in Section 3.3, the prediction of the heat transfer and pressure loss characteristics of 3D CFD simulations are used in this thesis for the development of heat sink correlations. These correlations can then be integrated into simplified 1D models to enable accurate prediction of flow characteristics with reduced computational effort. For this purpose, the heat sink is still analyzed in isolation. Based on the previous results demonstrating the negligible impact of stacking multiple channels, a single wavy mini-channel is used for the correlation development.

Since correlations are only valid within the range of the input data, a sweep of Reynolds and Prandtl numbers is performed to cover the expected operating range of the BTMS. The single mini-channel Reynolds number is varied as $Re = 50, 150, 200, 250, 350$ and the Prandtl number as $Pr = 25.9, 15.8$. These conditions correspond for the chosen coolant (EGL 50:50) to BTMS cooling circuit volume flow rates of $8 - 94\,{}^{\mathrm{l}}/_{\mathrm{min}}$ and inlet temperatures of $T_{\mathrm{in}} = 25, 45\,^{\circ}\mathrm{C}$, based on the defined reference BTMS concept. The 3D CFD simulation setup assumes steady-state conditions, incompressible laminar flow, constant fluid properties and thermally as well as hydrodynamically developing flow. Since the correlation development focuses on 1D/3D model coupling, particular emphasis is placed on accurately capturing the local heat transfer characteristics experienced by each individual LIB cell. Therefore, the local Nusselt number is calculated over wavelength long channel segments, each corresponding to the heat transfer region associated with a single LIB cell.

The simulation results for the local Nusselt number are presented in the characteristic map shown in Figure 5.10. The results can be categorized based on the channel length. Notably, for all data sets, a non-constant local

Nusselt number is observed in the entrance region of the channel, specifically for lengths $y < 0.15$ m. This behavior is attributed to the hydrodynamic and thermal developing flow, which is expected in the entry region of the channel.

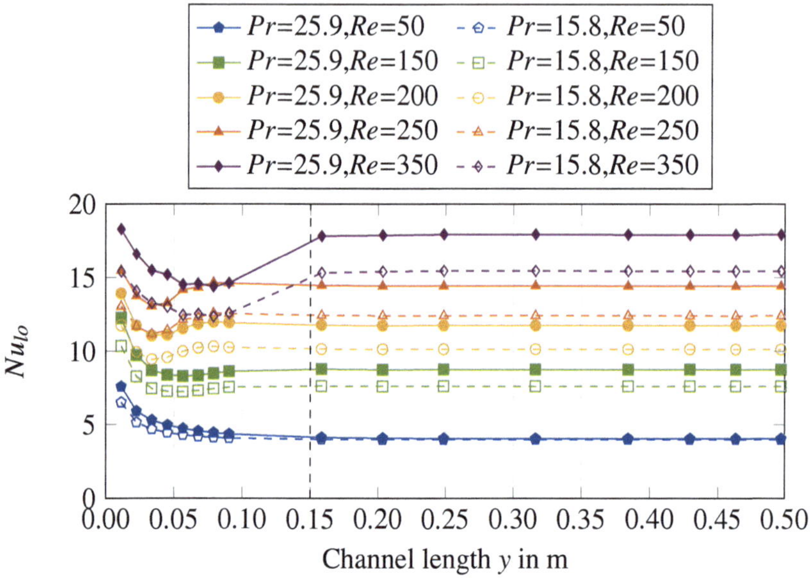

Figure 5.10: Local Nusselt number Nu_{lo} over channel length y for different Re and Pr. Nu_{lo} is calculated over wavelength long channel segments.

Beyond a channel length of $y = 0.15$ m, the local Nusselt number becomes nearly constant for all cases, except for $Re = 50$, where the flow remains similar to that in a straight channel due to the low fluid velocity. This constant behavior of the local Nusselt number is characteristic of laminar flow within rectangular wavy mini-channels. In such channels, the curvature introduced by the wavy design generates centrifugal forces that act on the fluid, inducing secondary flow structures known as Dean vortices. These vortices play a crucial role in the thermal performance of the system in two ways. First, they stabilize the local heat transfer characteristics, as evidenced by the observed constant Nusselt number after the entry region. Second, they enhance fluid mixing across the channel cross-section, thereby improving heat transfer relative to a straight channel designs. [114]

However, this enhanced mixing also leads to increased pressure losses compared to straight channels. As the focus of this study is on the development of a CFD-based empirical correlation, a detailed analysis of the underlying flow phenomena is beyond its scope. For a comprehensive discussion of the secondary flow behavior in wavy mini-channels, the reader is referred to Sui et al. [114].

Taking into account the total length of the reference battery module (approximately 1.74 m), the observed constant local Nusselt number beyond a channel length of 0.15 m can be utilized. As a result, the CFD-derived local Nusselt number at the axial channel position $y = 0.5$ m is selected for the development of the correlation. Using these values, a parametric curve-fitting approach is applied, which is shown in Figure 5.11 a).

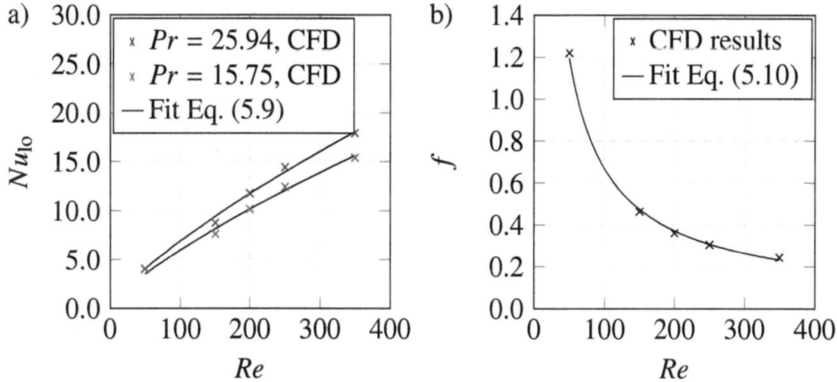

Figure 5.11: Parametric curve fitting applied for: a) Local Nusselt number Nu_{lo} and b) Darcy friction factor f.

For this purpose, the structure of the correlation given in Equation (5.8) is used. Since the geometry of the heat sink design is not changed in this thesis, the correlation results as

$$Nu_{lo}(Re, Pr) = 0.0785\, Re^{0.7675}\, Pr^{0.2895} \tag{5.9}$$

dependent on the Reynolds Re and Prandtl Pr number. The correlation demonstrates good agreement with the CFD data, exhibiting a maximum deviation of 13 % at $Re = 50, Pr = 15.8$ and a mean absolute deviation across all evaluated operating conditions of 3.6 %.

Similar to the Nusselt number correlation, the CFD results are also used to derive a pressure loss correlation in the form of the Darcy friction factor, as shown in Figure 5.11 b). The results indicate that the pressure loss is independent of the Prandtl number and depends solely on the Reynolds number of the flow. Again using a parametric curve-fitting approach, the following correlation for the Darcy friction factor is obtained

$$f(Re) = 32.0971\, Re^{-0.8412} \tag{5.10}$$

depending only on the Reynolds number. The derived correlation shows good agreement with the CFD data, exhibiting a maximum deviation of 4.4 % at $Re = 300$ and a mean absolute deviation of 2.6 %.

Both proposed correlations are valid within the investigated parameter range of $Re = 50-350$, $Pr = 15.8-25.9$, using EGL 50:50 as the coolant and for the specified rectangular wavy mini-channel geometry. These correlations are implemented for the heat sink in the 1D module section and module-level models, enabling accurate simulation of the electro-thermal-fluid behavior of the battery module under applied cooling conditions.

6 Results

Building on the previously validated cell-level models and the CFD-predicted characteristics of the heat sink, this chapter presents the module section and full module-level results. The module section includes a total of 24 LIB cells, the wavy mini-channel heat sink and the structural foam between all components, as described in Section 4.3. Since no direct experimental validation data is available at the module section or full module level, the 3D module section simulation is used to evaluate the accuracy of the developed 1D model. Consequently, the 3D and 1D mean jelly roll temperature predictions are compared with special emphasis placed on capturing cell-to-cell temperature differences. By scaling up from the module section to the full module level, finally, the proposed BTMS concepts can be evaluated and compared with respect to their temperature-related EOL aging characteristics.

6.1 Module section level results

The module section level results comprise the quantification of cell-to-cell temperature inhomogeneities and the comparison between the more detailed 3D model and the lumped 1D model. For this purpose, the previously derived Nusselt number and pressure loss correlations are implemented within the thermal-hydraulic pipe flow component of the 1D model to accurately capture the predicted heat transfer behavior. The results are discussed in the following sections.

Load profile

For the application of the 3D/1D approach, the fast charge scenario from the cell level validation is used, as illustrated in Figure 5.4 d). This profile is selected because a demanding load condition is necessary to induce noticeable cell-to-cell temperature differences under active cooling. Fast charging represents the most thermally challenging operating condition for a BEV and is therefore particularly suited for this purpose.

M. Auch, *Holistic Modeling Framework for the Investigation of Innovative Battery Thermal Management Concepts*, Wissenschaftliche Reihe Fahrzeugtechnik Universität Stuttgart, https://doi.org/10.1007/978-3-658-52612-2_6

Furthermore, the boundary conditions for the heat sink are defined for each single mini-channel with an inlet velocity of $v_{in} = 0.29\,{}^{m}/_{s}$, resulting in an inlet Reynolds number of $Re = 200$ and an inlet temperature of $T_{in} = 25\,°C$. In addition, temperature-dependent fluid properties are used to accurately capture the changing heat transfer conditions resulting from variations in the coolant's thermal properties. As illustrated in Figure 6.1, the cell numbering is performed in ascending order based on the cells' positions along the axial length of the cooling channel, starting with Cell 1 at the inlet and ending with Cell 24 at the outlet.

Figure 6.1: Module section level cell numbering.

Temperature inhomogeneities at both the cell level and across all cells at the module level play a crucial role in BTMS design. The 3D simulation results, based on the MPH-TR cell-level modeling approach and illustrated in Figure 6.2, provide detailed insights into the temperature distribution within individual cells along the module length. Although the results indicate that, compared to the laboratory thermal boundary conditions shown in Appendix Figure E.1, the temperature gradient shifts from the cell length direction to the radial direction due to the applied side cooling approach, this aspect is not further investigated in this work. Instead, this thesis focuses on capturing the cell-to-cell mean jelly roll temperature differences that drive inhomogeneous aging phenomena at the module level.

Temperature level

Figure 6.3 presents a comparison of the overall temperature predictions during the applied load profile between the 3D and 1D module section models. For this purpose, the volume-averaged jelly roll temperatures $T_{avg,jr}$ of the first (Cell 1), eighth (Cell 8), sixteenth (Cell 16) and last (Cell 24) cells are shown. The 3D and 1D models exhibit good agreement in both the absolute temperature levels and the temperature evolution over the

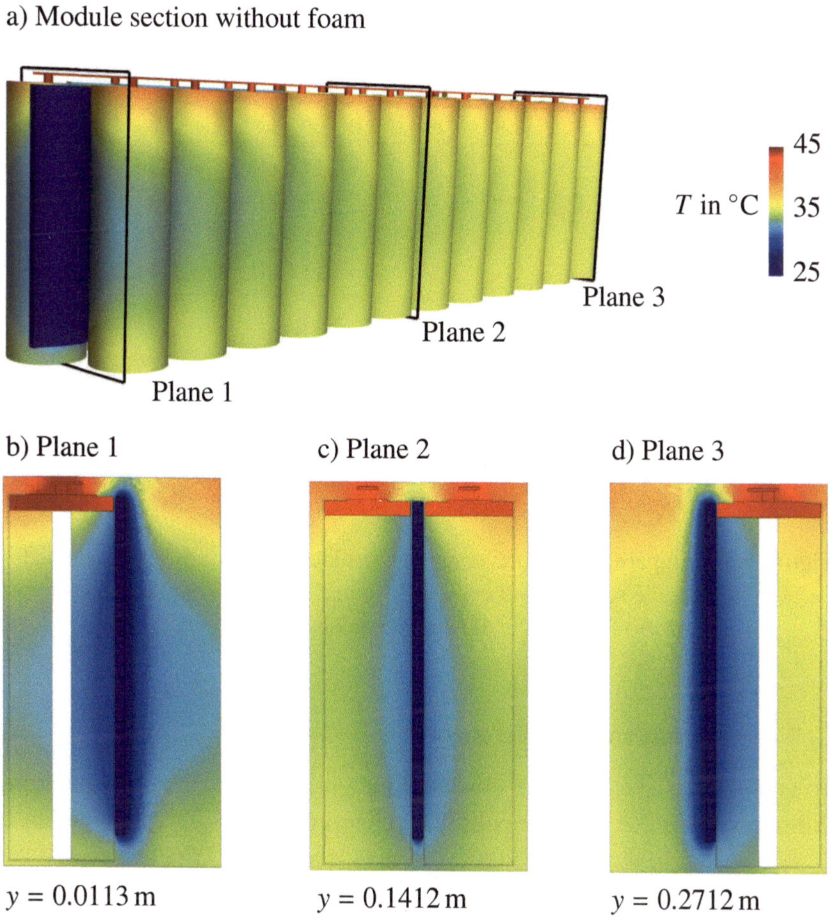

Figure 6.2: Module section temperature during the studied fast charge scenario at $t = 630\,\text{s}$: a) Complete section without foam, b) Plane in the center of Cell 1, c) Plane in the center of the module section and d) Plane in the center of Cell 24.

fast charge scenario for all four analyzed cells. Using the RMSE as the evaluation metric between the two frameworks, the deviations are calculated as $\text{RMSE}_{\text{Cell1}} = 0.21\,\text{K}$, $\text{RMSE}_{\text{Cell8}} = 0.37\,\text{K}$, $\text{RMSE}_{\text{Cell16}} = 0.43\,\text{K}$

and $\text{RMSE}_{\text{Cell24}} = 0.44\,\text{K}$, all remaining below 0.5 K. However, a trend of increasing RMSE with increasing cell number can be observed.

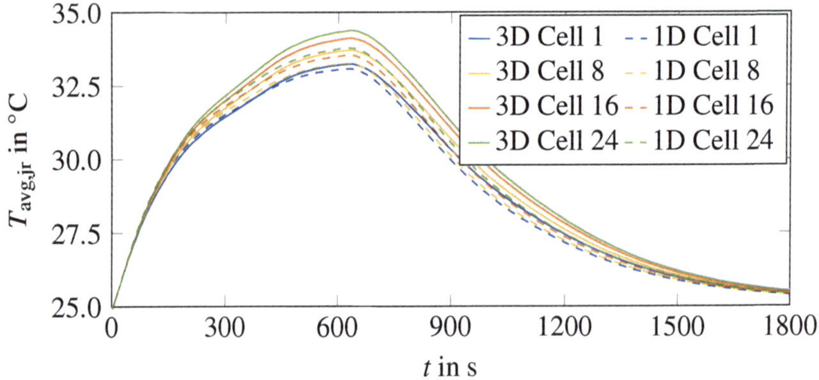

Figure 6.3: Comparison of the average jelly roll temperature $T_{\text{avg,jr}}$ of selected module section level cells 1, 8, 16 and 24 in the simulated fast charge scenario for both the 3D and 1D models.

Cell-to-cell temperature differences

To analyze the source of the observed increasing RMSE, the temperature difference averaged over the entire fast charging scenario is calculated by

$$\Delta T_{\text{avg},i,j} = \frac{\sum_{t=0}^{t_{\text{end}}} \left(T_{\text{avg,jr},j}(t) - T_{\text{avg,jr},i}(t) \right)}{t_{\text{end}}} \tag{6.1}$$

where i and j represent two neighboring cells, with $j = i + 1$. The results are illustrated in Figure 6.4.

In total, there are three different results shown, the 3D simulation results including the busbar, neglecting the busbar (where the thermal properties of the busbar are replaced with those of the foam) and the 1D model. Comparing the 3D simulation results, a significant difference in the cell-to-cell temperature predictions can be identified between the cases with and without the busbar, although the additional ohmic heat generation from the electrical resistance of the busbar is neglected in this study. The most pronounced

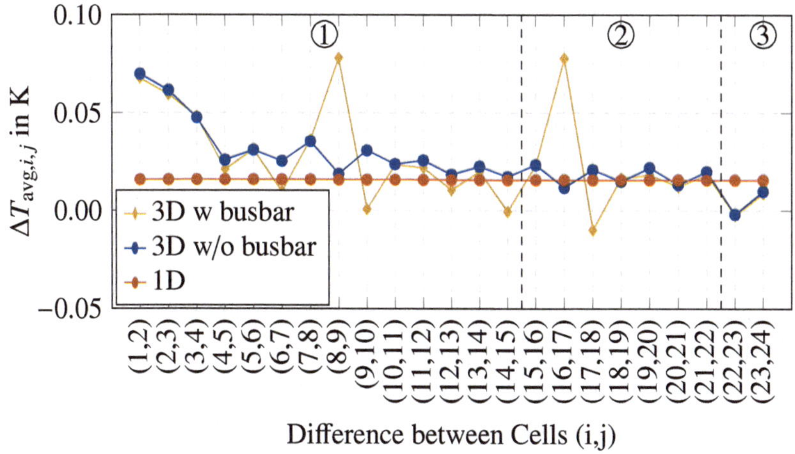

Figure 6.4: Average temperature difference $\Delta T_{\text{avg}i,j}$ between neighboring cells (i, j) over the simulated fast charge scenario. Three regions: 1) Entry length region, 2) Constant temperature difference region and 3) Exit length region.

differences occur between the cell pairs (8,9) and (16,17). These cell pairs represent the series busbar connection to the subsequent group of eight parallel-connected cells. This observation highlights that additional thermal inhomogeneities may arise solely from the busbar material and its non-symmetric design, which is caused by the coexistence of parallel and series connections within the module. However, since evaluating specific busbar designs is beyond the scope of this thesis and the maximum temperature difference within the module section is not significantly changed, the busbar is neglected in the subsequent analysis. This approach is consistent with the 1D model, where the busbar is also not considered.

Based on the 3D results excluding the busbar, the module section can be divided into three distinct regions, which are 1) the entry-length region, 2) the constant temperature-difference region and 3) the exit-length region.

The entry-length region is defined based on its corresponding segment of the cooling channel. This region spans a total channel length of approximately 0.157 m, extending from the channel inlet to Cell 15. As shown in Figure 5.10, the behavior of the local Nusselt number Nu_{lo} provides insight into the observed decrease in temperature differences between neighboring cells within this region. Notably, the local Nusselt number does not reach a constant value. Instead, it initially decreases and then increases. As a result, the heat transfer characteristics vary significantly from cell to cell in the entry region. This spatial variation is not captured by the Nusselt correlation used in the 1D simulation, which assumes a fully developed flow. Consequently, the 1D model is not expected to accurately represent the heat transfer behavior in the entry-length region.

These findings further explain the deviations observed in the average jelly roll temperatures, as illustrated in Figure 6.3. In the entry-length region, the use of the constant Nusselt correlation in the 1D simulation results in an overestimation of heat transfer. Consequently, the predicted average jelly roll temperatures for Cells 8, 16 and 24 are lower in the 1D simulation compared to the 3D simulation. Despite this discrepancy, the RMSE values for these cells do not increase significantly. This deviation is both expected and acceptable, given that the 1D model applies a simplified, constant local Nusselt number rather than a more detailed, position-dependent correlation.

The second region, defined by the neighboring cell pairs from (15,16) to (21,22), is characterized by an almost constant temperature difference between adjacent cells. This uniformity is a result of the local Nusselt number stabilizing beyond a cooling channel length of 0.15 m. In this region, the 3D and 1D simulation results show very good agreement. The average temperature difference between neighboring cells is $\Delta T_{3D,avg,i,j} = 0.018\,K$ in the 3D simulation and $\Delta T_{1D,avg,i,j} = 0.016\,K$ in the 1D model.

In the third region, the temperature difference between neighboring cells is influenced by exit length effects. Unlike the entry region, this variation is not caused by changes in the coolant flow, but rather by the absence of downstream cells beyond the final two (Cells 23 and 24). As a result, there is no additional heat input from subsequent cells. This effect lowers the average jelly roll temperatures of Cells 23 and 24, thereby decreasing the temperature difference relative to their preceding neighboring cells.

Finally, the pressure drop predicted by the 1D and 3D simulations during the fast charge event is compared. The 3D CFD model yields a pressure drop of $\Delta p_{3D} = 2911\,\text{Pa}$, while the 1D model predicts $\Delta p_{1D} = 3100\,\text{Pa}$ using the derived Darcy friction factor correlation given in Equation (5.10). This corresponds to a deviation of 6.5 %, demonstrating good agreement between both modeling approaches and confirming the accuracy of the derived friction factor correlation.

Although the constant temperature difference region (Region 2)) appears narrow compared to the entry-length region (Region 1)), this perception changes when considering the full module. At the full module level, a total of 154 cells are cooled by a single heat sink. Excluding the first 15 cells and the final two, the constant temperature difference region effectively encompasses approximately 90 % of the cells. As a result, the derived local Nusselt number and friction factor correlations are considered valid for accurately capturing the heat transfer and pressure loss characteristics within the cooling channel. This enables the scalability of both the battery cell count and the cooling channel length from the module section to the full module geometry for the 1D model.

6.2 Full module level results

In the following, the full module-level BTMS concepts are evaluated, with a particular focus on their life-cycle performance in relation to temperature-dependent aging behavior. For this purpose, the previously validated aging model for capacity loss and resistance increase is incorporated into the electrical LIB cell model. Furthermore, the upscaling from the module-section level to the full module level, as presented in Section 4.4, is applied. This results in the model reduction of investigating the 154 cells cooled by a single heat sink, which are further simplified into 11 thermal bricks, each representing 14 LIB cells, as illustrated in Figure 6.5.

As a result of this approach, the temperature difference between the first and last bricks underestimates the true temperature gradient across the first and last LIB cells. However, this introduces a conservative assumption when assessing the potential temperature-related inhomogeneous aging reduction

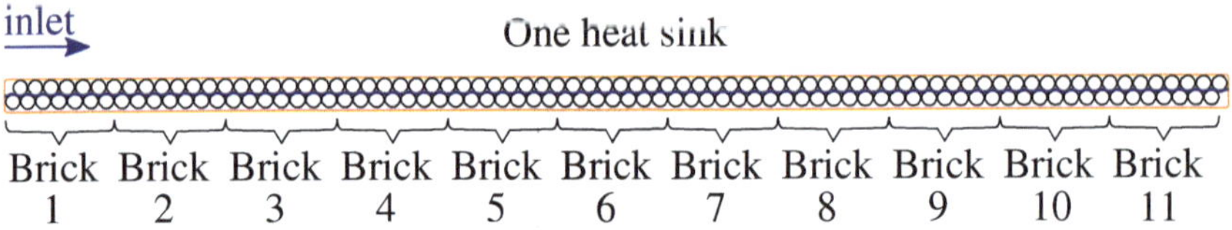

Figure 6.5: Model simplification for one heat sink from individual LIB cells to brick level (each brick consisting of 14 LIB cells). Complete model reduction process is illustrated in Figure 4.11.

provided by the proposed switchable BTMS concept. Regardless of the investigated BTMS concept, the brick numbering remains consistent and is defined in reference to the conventional unidirectional BTMS concept. Brick 1 corresponds to the front of the heat sink, while Brick 11 represents the rear. For the investigation, the life-cycle study load profile is first defined, followed by an evaluation of the conventional unidirectional, switchable and two additional advanced BTMS concepts (split-cooling and U-turn channel design) with respect to their EOL characteristics.

6.2.1 Load profile

Since customer expectations for BEVs include achieving charging times comparable to the refueling times of ICE vehicles, it is assumed in this study that exclusively fast charging is applied. As the LIB cell's maximum charging C-rate and temperature characteristics constrain the maximum charging power, the load profile is defined from the cell level upwards. The resulting cell-level current and voltage profile is illustrated in Figure 6.6 under the assumption that the charging station can supply sufficient power.

The definition of each load profile step is summarized in Table 6.1. As shown in Section 5.1.2, drive cycles such as the WLTC scenario do not result in significant temperature rises for the LIB cells in the studied reference BTMS design. Consequently, the driving scenario (discharging) is simplified as a constant current discharge at a 0.5 C rate. In actual BEVs, the battery management system aims to limit the SOC range to avoid operation in the upper and lower extremes, as these regions lead to accelerated aging and increased heat generation [99]. Therefore, for the load profile in this study,

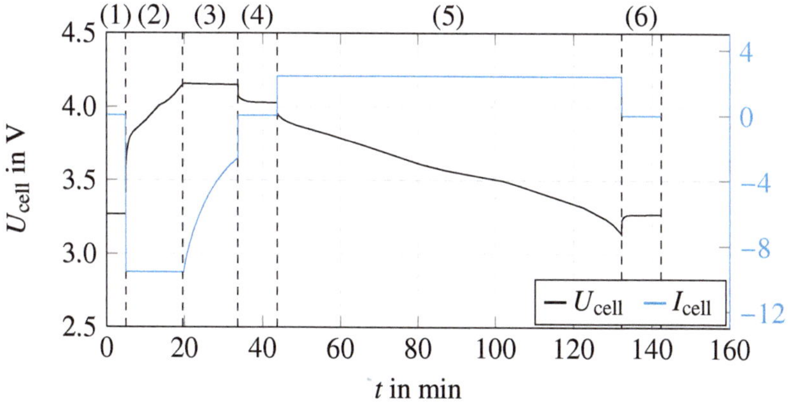

Figure 6.6: Voltage and current of the defined life-cycle load profile. Black dashed lines indicate load profile step changes shown in Table 6.1.

the CC-CV charging scenario is restricted to a voltage of 4.15 V, while the discharging scenario is limited to a voltage of 3 V, which corresponds to the voltage range chosen for the previously shown experimental aging study. At BOL the initial SOC is set at 5 %. The pause phases are included to enable thermal relaxation between major load steps, ensuring consistent starting temperatures for the fast charging and discharging phases throughout the life-cycle. As no current is applied during these phases, they do not contribute to the aging of the LIB cells.

Table 6.1: Defined load profile for the life-cycle study.

Step	Name	Boundary Conditions	
(1)	Pre-Conditioning	$T_{in} = 25\,°C$,	end if $t_{step} = 300\,s$
(2)	Fast-Charge CC	$I_{cell} = -9.6\,A$,	end if $U_{cell} = 4.15\,V$
(3)	Fast-Charge CV	$U_{cell} = 4.15\,V$,	end if SOC $\geq 80\,\%$
(4)	Pause	$I_{cell} = 0\,A$,	end if $t_{step} = 600\,s$
(5)	Drive CC	$I_{cell} = 2.4\,A$,	end if $U_{cell} = 3\,V$
(6)	Pause	$I_{cell} = 0\,A$,	end if $t_{step} = 600\,s$

Regarding the BTMS system boundary conditions, a constant inlet temperature of the coolant at T_{in} = 25 °C is assumed over the entire aging study. This inlet temperature ensures that no significant temperature gradients are induced within the LIB cells due to excessively large temperature differences between the coolant and the cells. Furthermore, the BTMS volume flow rate is iteratively determined as 53.2 l/min corresponding to one stacked wavy mini-channel flow rate of 1.9 l/min, as illustrated in Figure 6.7.

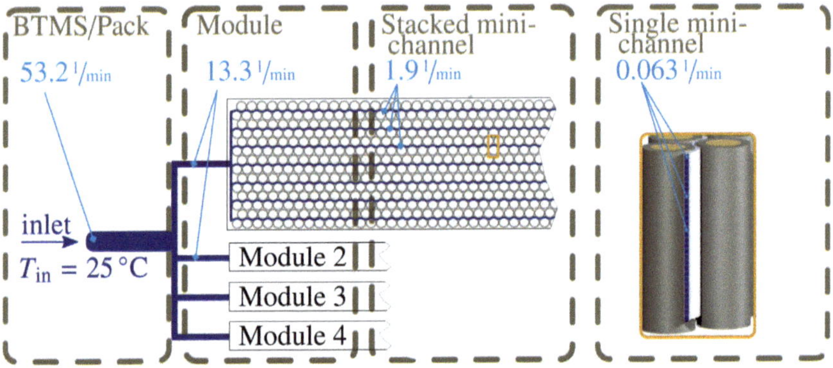

Figure 6.7: Schematic of the volume flow rate distribution from BTMS to single wavy mini channel.

This boundary conditions are chosen to meet two key thermal requirements. Firstly, the mean jelly roll temperature of all bricks remains below 40 °C, fulfilling a safety margin relative to the maximum charging temperature $T_{max,cell,charge}$ = 45 °C specified in the data sheet [82]. Secondly, the state-of-the-art accepted cell-to-cell temperature difference of 5 K between cells within the module should not be exceeded during the entire life-cycle study for the conventional BTMS concept.

6.2.2 Conventional unidirectional BTMS concept

As baseline concept, the conventional BTMS concept shown in Figure 3.1 is investigated first. This concept is characterized by an unidirectional flow through the heat sink, with the inlet and outlet located on opposite sides of the battery module throughout the entire life cycle. The full module level simulations are run until each brick reaches the EOL criterion set as a SOH of 80 %. The results of the conventional BTMS concept are additionally used to ensure the fulfillment of the set requirements on the maximum average jelly roll temperatures and cell-to-cell temperature differences within the module. The results for bricks 1,6 and 11 are presented in Table 6.2 and both temperature requirements are fulfilled.

Table 6.2: Life-cycle study results of the conventional BTMS concept.

Brick	$T_{max,jr}$ in °C at BOL	$T_{max,jr}$ in °C at EOL	SOH in % at EOL	Charge throughput in Ah until EOL_{Brick}
1	33.74	34.66	81.68	5888
6	35.95	37.05	80.85	5476
11	38.10	39.35	80.00	5100
Δ_{max}	4.36	4.69	1.68	788

The mean jelly roll temperature is below 40 °C at both BOL and when the first brick reaches EOL. The requirement for the maximum cell-to-cell temperature difference within the module is also fulfilled with 4.36 K at BOL and 4.69 K at EOL. The remaining margin to the 5 K limit is considered, since the bricks summarize the mean jelly roll temperature of 14 LIB cells, consequently underestimating the actual temperature difference within the module. As previously explained, this temperature difference arises due to the coolant absorbing heat from the LIB cells and heating up downstream the cooling channel. The observed maximum temperature and cell-to-cell temperature difference increase over the life-cycle is due to the aging-related increased resistance of the LIB cells during the life-cycle study. In Figure 6.8 a) and c), the mean jelly roll temperature and SOH over the first four load cycles are illustrated. Figures 6.8 b) and d) present the corresponding results for the switchable BTMS concept, which will be discussed in the following section and is included here to facilitate comparison between the two concepts.

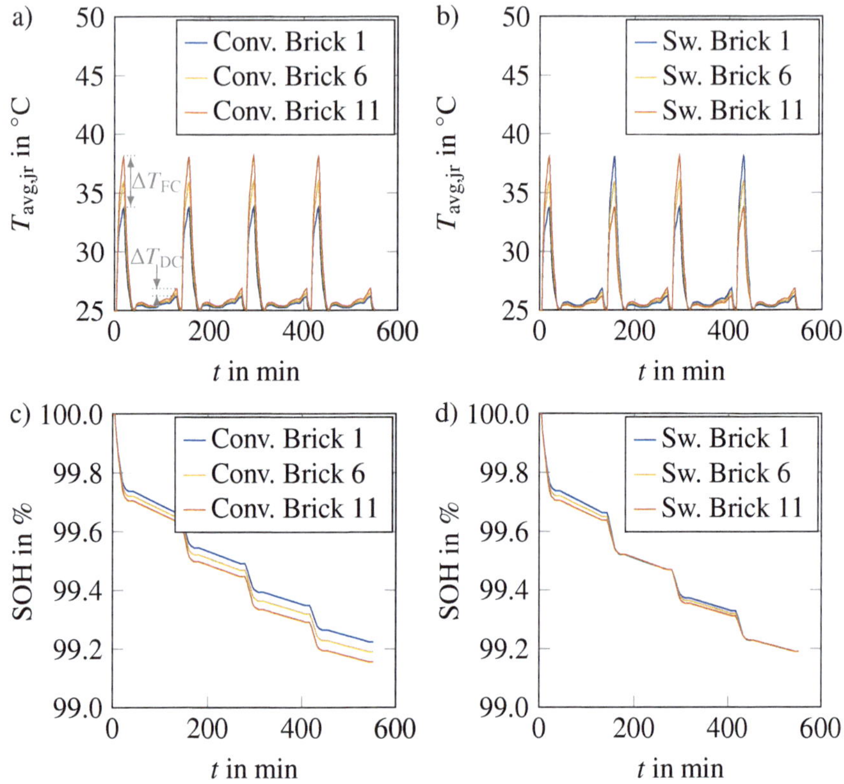

Figure 6.8: Average jelly roll temperature $T_{avg,jr}$ for bricks 1,6,11 in: a) Conventional, b) Switchable BTMS concept and SOH in: c) Conventional and d) Switchable BTMS concept during the first four cycles.

In the conventional BTMS concept, the brick located at the inlet (Brick 1) is characterized by the lowest mean jelly roll temperature and as a consequence, highest SOH over all four load cycles. In comparison, the brick located at the rear (Brick 11) exhibits the highest mean jelly roll temperature and as a result, lowest SOH over the first load cycles. The brick located in the center of the module (Brick 6) is in between the outer bricks in both the mean jelly roll temperature and the SOH behavior. As shown in

Figure 6.8 a), the highest temperature level and cell-to-cell temperature differences (ΔT_{FC} = 4.36 K) can be identified during the fast charging. In comparison, during the discharge, almost a magnitude lower temperature difference (ΔT_{DC} = 0.64 K) is observed. As a consequence, the accelerated aging can be identified in Figure 6.8 c) especially during the fast charging phases of the load profile. The unidirectional flow of the conventional BTMS concept leads to always the same cells being exposed to the highest temperatures. This results in an increasing SOH difference with increasing count of load cycles. Finally, the SOH is shown over the entire life-cycle of the conventional BTMS concept in Figure 6.9 a). Consistent with the previous figure, Figure 6.9 b) is included to provide the results for the switchable BTMS concept, which will be discussed in the subsequent section.

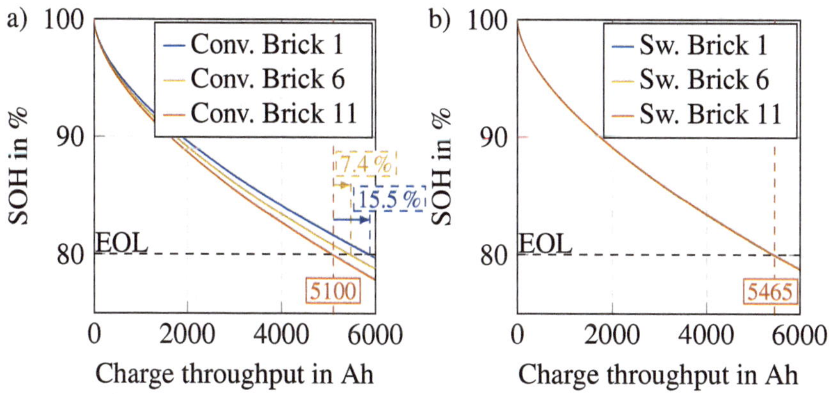

Figure 6.9: SOH over charge throughput for the entire life-cycle study in: a) Conventional and b) Switchable BTMS concept.

Although active balancing can mitigate the influence on BEV performance of inhomogeneously aged cells, it cannot restore the lost capacity of accelerated-aged LIB cells. As a consequence, the weakest cell determines the EOL of the entire pack [72, 115]. As shown in Table 6.2, Brick 11 reaches the EOL criterion of SOH = 80 % first, with a total charge throughput of 5100 Ah (≈ 285000 km cumulative mileage over the life-cycle). This reduced charge throughput, compared to the other bricks, is attributed to the previously discussed highest operating temperature during the load profile. At the time step when Brick 11 reaches its EOL, Brick 6 (80.85 %) and Brick 1 (81.68 %) would still be considered to not have reached their respective EOL. Although this SOH difference appears minor, due to the sublinear aging behavior, it has a significant impact on the remaining possible charge throughput, assuming that the knee point (sudden increase of aging rate) is reached below the defined EOL limit. As illustrated in Figure 6.9 a), the charge throughput until their respective EOL for Brick 6 and Brick 1 is increased by 7.4 % (5476 Ah) and 15.5 % (5888 Ah), respectively. Consequently, there is a significant potential to increase the total charge throughput of the battery pack by reducing cumulative temperature-related aging inhomogeneities.

6.2.3 Switchable BTMS concept

With the temperature-related inhomogeneous aging influence on the battery pack life-cycle characteristics of the conventional BTMS with unidirectional flow quantified, the potential improvements by the switchable BTMS concept introduced in Section 3.2 can be discussed. As illustrated in Figure 3.3, the switchable BTMS concept is realized by the integration of an additional 4/2 valve, which enables to alter the flow direction. The switch is activated in the simulation model through a counter, which increases by one every time a fast charge scenario is applied on the battery module. During vehicle standstill after the fast charge is finished (corresponding to step (4) of the load profile in Table 6.1), the counter activates the switch to alter the flow direction. The built 1D simulation model including the control system for the switchable BTMS concept is shown in Appendix Figure D.2. For this study, no additional pressure losses are considered by the 4/2 valve. As shown in Table 6.3, the temperature requirements for both the maximum temperature and cell-to-cell temperature difference are fulfilled over the entire life-cycle study.

Similar to the conventional BTMS concept, the mean jelly roll temperature and SOH over the first four load cycles are discussed first with Figure 6.8 b) and d). The mean jelly temperature behavior results prove the successful implementation of the switchable BTMS concept. As aimed for,

Table 6.3: Life-cycle analysis results of the switchable BTMS concept.

Brick	$T_{max,jr}$ in °C at BOL	$T_{max,jr}$ in °C at EOL	SOH in % at EOL	Charge throughput in Ah until EOL_{Brick}
1	33.74	39.28	80.00	5466
6	35.95	37.08	80.02	5475
11	38.10	34.77	80.00	5465
Δ_{max}	4.36	4.51	0.02	10

in every subsequent cycle the previously hottest brick become the coldest brick and vice versa. The brick located in the center of the module keeps its temperature level over the subsequent load cycles. The achieved changing average jelly roll temperatures of the switchable BTMS concept have a significant impact on the SOH behavior of the bricks.

While for the first fast charge, the SOH behavior is similar to the conventional BTMS concept, for the second fast charging scenario the effect of altering the flow direction shows its benefits. As a consequence of the changed temperature levels between Brick 1 and 11, the previously fastest aged Brick 11 ages the slowest in the subsequent fast charge and vice versa for the previously coldest Brick 1. Since Brick 6 remains on its temperature level, this leads to all bricks being back to similar SOHs after two load profile cycles. This effect can be identically observed for the subsequent two load profile cycles. As illustrated in Figure 6.9 b), the switchable BTMS balances the temperature-related aging inhomogeneities over the entire battery pack life-cycle.

Visually, the charge throughput dependent SOH behavior of the bricks can not be differentiated. By using the life-cycle characteristics given in Table 6.3, Brick 1 and 11 reach the EOL criterion as the first bricks at the same time step and a difference to the SOH of Brick 6 of only 0.02 % can be identified. The maximum charge throughput difference until each brick reaches its respective EOL is only at 0.18 % (10 Ah).

While the switchable BTMS concept fulfills the goal of reducing the cumulative temperature-related aging inhomogeneities compared to the conventional unidirectional BTMS concept, the improvement in total charge throughput until the first brick reaches EOL needs to be quantified. The conventional BTMS concept reaches the EOL with a total charge throughput of 5100 Ah, while for the switchable BTMS concept the EOL is reached at a charge throughput of 5465 Ah. Consequently, the proposed concept not only homogenizes the aging behavior but also improves the total charge throughput by 7.2 % (365 Ah). This increase directly translates to a 7.2 % ($\approx$ 20000 km) increase in cumulative mileage of the BEV until the EOL is reached. The additional pressure penalty in this study is zero, due to the neglection of additional pressure losses induced by the integration of the 4/2 directional valve.

While switching after every cycle provides the greatest benefit, it requires accurate detection of each fast charge scenario of the BEV. To evaluate the potential impact of inaccurate detection, wider switching intervals are analyzed and compared with respect to their EOL characteristics.

6.2.4 Investigation of switch interval

In addition to the previously discussed 1 cycle switch interval, intervals of 5, 10, 25 and 50 cycle switch intervals are investigated. For this purpose, the overall switchable BTMS concept architecture remains the same and the switch is activated after the fast charge scenario counter reaches the respective switch interval value. The life-cycle characteristic results are shown in Table 6.4.

Using the conventional BTMS concept as benchmark, the results indicate that the charge throughput increase benefits of the switchable BTMS concepts are similar for switching intervals of 1 (7.16 %), 5 (7.14 %) and 10 (7.14 %). While further increasing the switch interval still shows significant improvements compared to the conventional BTMS concept, the effect is slightly reduced to a 6.98 % and 6.96 % for the 25 and 50 cycle switch interval, respectively. The reduced effect can be further analyzed using Figure 6.10, where the maximum SOH difference SOH_{max} among all bricks is illustrated over the charge throughout for the investigated BTMS concepts.

Table 6.4: Life-cycle analysis results of the conventional and various switchable BTMS concepts.

Switch strategy	Charge throughput in Ah until EOL	ΔSOH_{max} bricks in %	Ah increase in % compared to conventional
Conventional	5100	1.68	0
1 cycle	5465	0.02	7.16
5 cycle	5464	0.02	7.14
10 cycle	5464	0.02	7.14
25 cycle	5456	0.05	6.98
50 cycle	5455	0.05	6.96

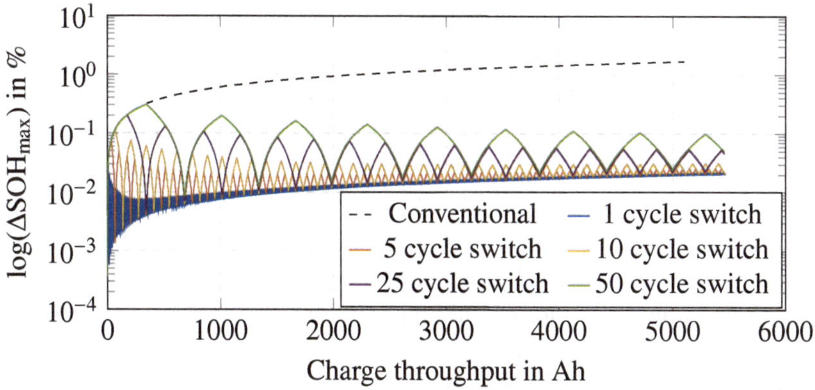

Figure 6.10: Maximum SOH difference ΔSOH_{max} between bricks for conventional and switchable BTMS concepts. Data cut off, when the first brick reaches EOL criterion.

The first observation is, that for the conventional BTMS concept the SOH difference increases monotonically over the charge throughput. This is due to the previously discussed effect of always the same bricks experiencing the higher operating temperatures. In contrary, for all switchable BTMS concepts, the SOH difference experiences a periodic oscillation behavior. Both the overall magnitude and the frequency is dependent on the switch interval. There are two key observations regarding the influence of the switch interval on the life-cycle characteristics.

First, increased switching intervals lead to temporary increased SOH differences due to the larger time frame in which the same brick experiences the higher operating temperature. Although the SOH difference decreases afterwards, this effect makes the potential improvements of the switchable BTMS concept dependent on the exact time step when the first brick reaches EOL. The higher the switching interval, the higher the risk that EOL is reached at a peak of the SOH difference. For the reference battery pack investigated in this thesis, using the 25 cycle switch interval the EOL is reached almost at the peak. Regarding the 50 cycle switch interval, the SOH difference is already halfway decreased, resulting in similar EOL characteristics of the battery pack for both switch cases. For narrower switch intervals, the SOH difference peaks are lower, placing less relevance on when the EOL criterion is reached.

The second key characteristic is that independent of the switch interval, always a SOH difference remains, which increases over the charge throughput. This effect is attributed to the assumed Arrhenius based aging model approach given in Equation (4.8). The capacity loss is proportional to the temperature within an exponential function. As a consequence, the aging behavior of operating the LIB cells always at the same temperature is not equal to the aging behavior of only having the same mean operating temperature over a longer period. This effect translates to the brick located in the module center always experiencing similar temperatures, while the outer bricks periodically change between higher and lower temperatures. Due to the maximum temperature difference within the module being limited to 5 K, this effect does not have a significant influence on the EOL characteristics (0.02 % SOH difference for 1 cycle switch interval).

While the conventional reference battery pack BTMS concept with unidirectional flow is chosen as benchmark, there exist other advanced solutions to mitigate the temperature-related inhomogeneous aging inside battery packs. However, they require substantial changes to either the BTMS architecture or the cooling channel design. Consequently, to further benchmark the proposed switchable BTMS concept, a simple approach is used for these concepts using the developed 1D simulation module-level model.

6.2.5 Comparison to advanced BTMS concepts

As illustrated in Figure 6.11 a), the conventional design (used in the reference vehicle) employs a uniform flow direction through the heat sink channels. The two investigated advanced BTMS concepts consist of a half-split cooling approach (Figure 6.11 b)) and a U-turn heat sink design [116] (Figure 6.11 c)). For the split-cooling approach, the BTMS architecture is adjusted that the coolant inlet position is changed to the center of the battery module. Consequently, there are two outlets at each side of the heat sink. In the U-turn heat sink design, the stacked mini-channel heat sink is divided into two sets of channel packages. As shown in Figure 6.11 c), the upper channel package (15 mini-channels) serves as inlet, while the lower channel package (15 mini-channels) serves as outlet. Consequently, the coolant has to transition from the upper to the lower channels at the end of the channel length. This concept aims to reduce the cell-to-cell temperature inhomogeneity. However, depending on the LIB cell thermal characteristics, this can increase thermal gradients over the cell length within single cells.

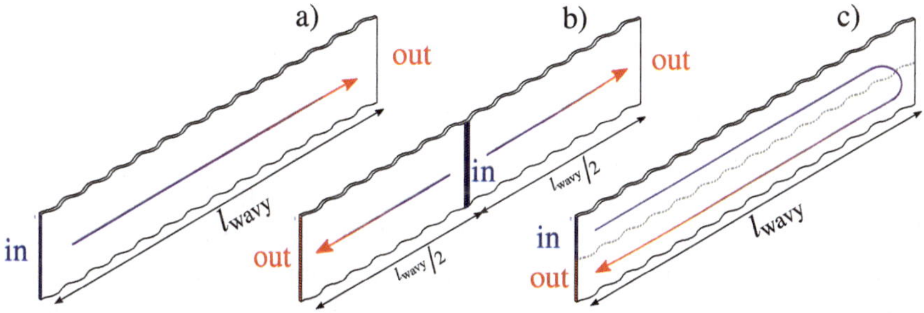

Figure 6.11: Schematic of different stacked wavy-mini-channel heat sink configurations: a) Conventional unidirectional, b) Half-split unidirectional and c) U-turn flow.

To ensure fair comparison between the conventional, switchable and advanced BTMS designs, the choice of boundary conditions is critical. Two approaches are possible. First, maintaining a constant total BTMS volumetric flow rate or second, keeping the flow rate in each single mini-channel constant. The latter would ensure a consistent single channel Reynolds number across all BTMS concepts but would lead to varying total BTMS coolant flow rates. This would result in different BTMS energy demands at the system level. Therefore, the first option is chosen, to keep the total BTMS coolant flow rate constant. This ensures vehicle-level comparability, although it results in concept dependent variations in the single channel Reynolds number and hence, single channel heat transfer characteristics.

Finally, the developed full module level BTMS model is slightly adapted for the investigation of the additional two advanced concepts. For the split cooling the adaption is integrated by splitting the brick located in the module center (Brick 6) into two smaller bricks, as illustrated in Figure 6.12. Consequently, the coolant inlet can be placed between those smaller bricks resulting in having 77 cells to be cooled in each direction.

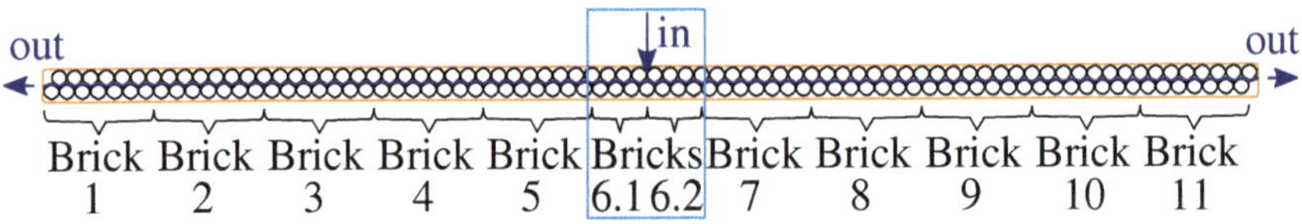

Figure 6.12: Adjusted model simplification for one heat sink from individual LIB cells to brick level units for the split cooling concept.

In the U-turn model, the stacked mini-channel heat sink is divided into two channel packages. As shown in Figure 6.11 c), the 15 upper channels serves as the inlet, where coolant enters and flows sequentially past each battery brick toward the rear end of the heat sink. At this point, the U-turn geometry redirects the coolant into the lower channels, enabling it to flow back along the bricks toward the outlet. Assuming a uniform temperature within the cell casing, each brick has a heat transfer path to both the upper and lower channel sections. The resulting brick temperatures over the first four load cycles for both advanced concepts are presented in Figure 6.13.

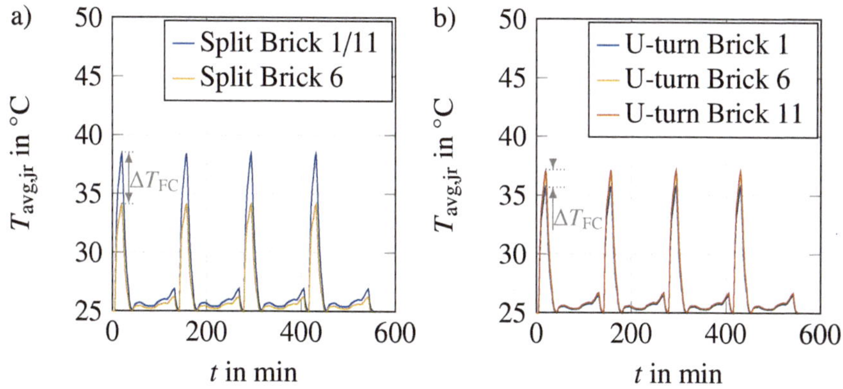

Figure 6.13: Average jelly roll temperature $T_{avg,jr}$ for bricks 1,6,11 during the first four cycles for: a) Split-cooling and b) U-turn.

A comparison of the average temperature results from the split-cooling approach (Figure 6.13 a)) with those of the conventional unidirectional BTMS concept (Figure 6.8 a)) reveals no significant difference in either the maximum temperature $T_{max,jr} = 38.28\,°C$ (Brick 1/11) or the cell-to-cell temperature difference $\Delta T_{FC} = 4.2\,K$. This outcome is primarily due to the split-cooling design halving the volumetric flow rate in each channel, thereby negating the potential benefit of having each coolant stream cool only half the number of LIB cells. As a result, the primary distinction introduced by the split-cooling approach lies in the distribution of temperature differences. Instead of a gradient from Brick 1 to Brick 11, the maximum cell-to-cell temperature differences now occur between both Brick 1-6 and Brick 6-11.

Regarding the average jelly roll temperatures of the U-turn BTMS concept shown in Figure 6.13 b), the maximum temperature difference within the module during the first load cycle is $\Delta T_{FC} = 1.4\,K$. This represents a 68 % reduction compared to the conventional unidirectional BTMS concept. While this is a significant improvement in thermal uniformity, the remaining temperature difference is primarily attributed to the non-uniform coolant temperature rise along the flow path. In the upper channel package, the temperature gradient between the LIB cells and the coolant is relatively large, enabling efficient heat absorption. However, toward the rear of the heat sink and along the return path through the lower channel package, this gradient decreases, limiting further heat absorption. As a result, the coolant temperature does not increase linearly along the flow path, leading to the observed remaining temperature difference within the module.

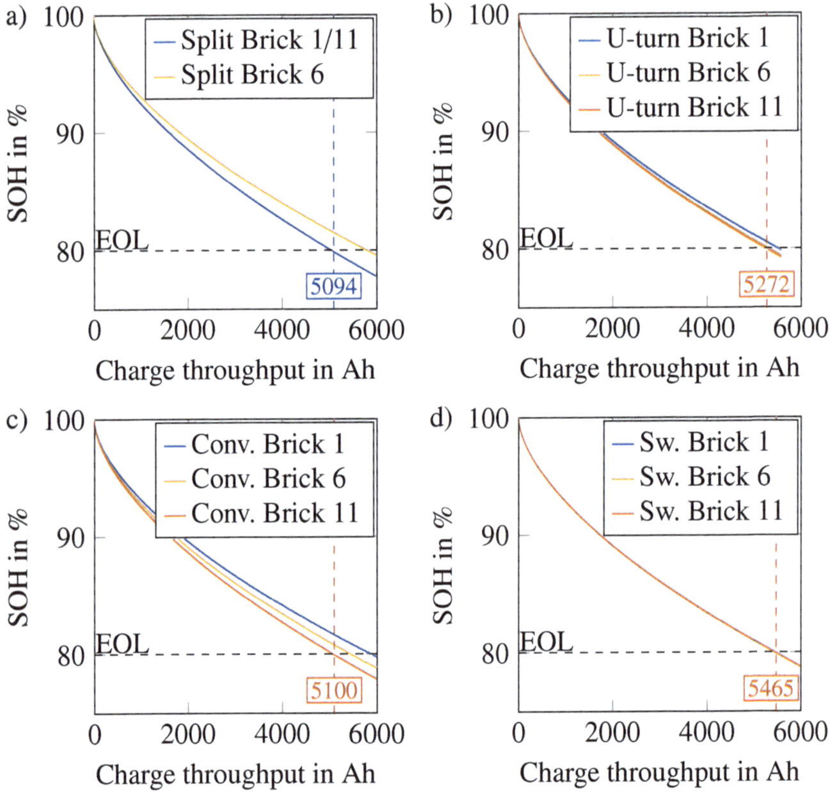

Figure 6.14: SOH over charge throughput for the life-cycle study for: a) Split-cooling, b) U-turn c) Conventional and d) Switchable BTMS.

Furthermore, this effect leads to a slightly reduced total removed heat across all cells compared to the conventional BTMS concept. Hence, the average temperature level of the bricks increases. The highest average jelly roll temperature is observed for Brick 11 at $T_{avg,jr}$ = 37.12 °C, while the lowest occurs for Brick 1 at $T_{avg,jr}$ = 35.72 °C. Compared to the conventional BTMS concept, the lowest brick temperature is increased by 5.87 % and reaches almost the level of the brick located at the center of the module in the conventional BTMS concept. Consistently to the conventional and switchable BTMS concepts, the described BTMS concepts are investigated until the EOL criterion is reached, as illustrated in Figure 6.14 a) and b).

As indicated by the only slightly increased average jelly roll temperatures of the bricks compared to the conventional BTMS concept and as shown in Table 6.5, the split-cooling approach results in a nearly identical EOL charge throughput, with a deviation of only -0.12 %. This minor difference is attributed to the slightly reduced heat transfer performance of the split-cooling configuration, caused by the lower Reynolds number in each mini-channel. However, the split-cooling approach reduces the pressure loss as indicated in Equation (2.26) by a factor of four, as both the channel length and the volumetric flow rate are halved.

Table 6.5: Life-cycle analysis results of the conventional, switchable, split-cooling and U-turn BTMS concepts.

Concept	Charge throughput in Ah until EOL	ΔSOH_{max} bricks in %	Ah increase in % compared to conventional	Pressure loss in Pa
Conventional	5100	1.68	0	18739
Sw. 1 cycle	5465	0.02	7.16	18738
Split-cooling	5094	1.65	-0.12	4578
U-turn	5272	0.56	3.37	83668

For the U-turn BTMS, the charge throughput until the EOL criterion is reached can be increased by 3.37 %. However, this improvement comes at the cost of an increase in pressure loss by a factor of four compared to the conventional BTMS concept, which is attributed to both the doubled volumetric flow rate and the extended flow channel length.

Among the evaluated BTMS concepts, the proposed switchable BTMS demonstrates the highest improvement in charge throughput until the EOL criterion is reached, as shown in Table 6.5 and illustrated in Appendix Figure F.1. While the split-cooling approach lowers the energy demand of the BTMS pump, it requires substantial modifications to the battery pack design and BTMS, particularly to integrate an additional coolant inlet at the center of the module. The U-turn heat sink design, on the other hand, increases manufacturing costs due to its more complex channel geometry and results in higher pump energy consumption.

In contrast, the integration of a 4/2 valve in the proposed switchable BTMS represents only a minor modification to the existing unidirectional BTMS architecture. Given that modern BEVs already include a connected VTMS valve system (as the Octovalve concept introduced in Figure 2.2), the addition of a single 4/2 valve can be implemented with minimal effort. This highlights the favorable trade-off between low realization cost and significant life-cycle performance improvement offered by the switchable BTMS concept.

7 Summary and conclusion

In this thesis, the development of cumulative temperature-induced aging inhomogeneities arising from the state-of-the-art unidirectional coolant flow of indirect liquid cooling BTMS concepts was identified as a key research gap. To tackle this challenge, a novel switchable BTMS concept is introduced that reverses the coolant flow direction after each complete load cycle during vehicle standstill by integrating a 4/2 valve into the existing BTMS architecture. Unlike existing BTMS strategies, this approach reduces cumulative temperature-related aging inhomogeneities over the life-cycle rather than focusing on improved temperature uniformity during operation.

A second major challenge was the lack of a reliable workflow to evaluate BTMS concepts over the full life-cycle while accounting for coupled electro-thermal-fluid-aging effects at the module level. To overcome this, a hierarchical thermal modeling approach was developed, ranging from individual cells to the full module. Compared to existing approaches, special focus was placed on accurately capturing BTMS-driven cell-to-cell temperature differences, while managing the associated computational effort. The approach results in a reliable and accurate coupled 1D electro-thermal-fluid-aging modeling framework, enabling efficient life-cycle assessment of BTMS concepts and facilitating adaptation to various system designs.

In the following, the implementation, validation, and results related to these contributions are summarized. At the cell level, improved ECM-coupled 1D and 3D thermal models were developed and validated against experimental data across multiple measurement points and load profiles, achieving RMSEs below 1 K. Additionally, an Arrhenius-based aging model was parameterized using empirical data to capture the dominant SEI-layer growth aging effect at operating temperatures of 25 °C and above.

As a benchmark, a reference BTMS design with unidirectional flow and a stacked wavy mini-channel heat sink was defined. Detailed 3D CFD simulations of the isolated heat sink were conducted to derive Nusselt number and Darcy friction factor correlations over the operating range. These correlations were integrated into the 1D module-level model to ensure accurate representation of heat transfer and pressure loss characteristics.

M. Auch, *Holistic Modeling Framework for the Investigation of Innovative Battery Thermal Management Concepts*, Wissenschaftliche Reihe Fahrzeugtechnik Universität Stuttgart, https://doi.org/10.1007/978-3-658-52612-2_7

To ensure the validity of the module-level 1D model, a hybrid 3D/1D approach was employed. A 24-cell section of the battery module was simulated with active cooling using a detailed 3D CFD model, which showed very good agreement with the corresponding 1D model predictions in terms of average jelly roll temperatures and cell-to-cell temperature differences. By integrating the validated aging model and scaling the module-section to full-module level, a quantitative comparison of the investigated BTMS concepts was conducted with respect to their EOL characteristics.

Compared to the conventional unidirectional BTMS design, the proposed switchable BTMS was found to increase the total charge throughput by 7.2 % until the EOL criterion (defined as the first cell reaching 80 % SOH) was met. This directly corresponds to a 7.2 % (≈ 20000 km) increase in cumulative mileage of the BEV until the EOL is reached. Among the evaluated switching intervals, changing flow direction after every complete load cycle yielded the most substantial benefits, though similar gains could still be achieved with switching intervals up to 10 cycles. The switchable BTMS concept also outperformed both the split-cooling and U-turn heat sink designs, further highlighting its practical relevance.

While in this thesis both the effectiveness of the switchable BTMS and the accuracy of the developed modeling framework was demonstrated, some limitations remain, suitable for future research. At the cell level, additional aging studies under varying C-rates and temperatures are needed to extend the applicability of the Arrhenius-based aging model. Moreover, integrating realistic busbar and cell contacting designs into the thermal model could improve the accuracy of the LIB cell's temperature level. Finally, experimental validation of the proposed BTMS concepts is recommended to reinforce the conclusions drawn from the simulations.

In conclusion, this work demonstrates the promising potential of the switchable BTMS concept to mitigate cumulative temperature-related aging inhomogeneities across the BEV life-cycle. The proposed solution is particularly interesting due to minor additional power requirements and architectural changes needed within existing VTMS and BTMS systems. As such, the switchable BTMS offers a cost-effective and practical concept to enhance battery life in future BEVs.

References

[1] SCHLEUSSNER, Carl-Friedrich ; ROGELJ, Joeri ; SCHAEFFER, Michiel ; LISSNER, Tabea ; LICKER, Rachel ; FISCHER, Erich M. ; KNUTTI, Reto ; LEVERMANN, Anders ; FRIELER, Katja ; HARE, William: Science and policy characteristics of the Paris Agreement temperature goal. In: *Nature climate change* 6 (2016), Nr. 9, S. 827–835. – DOI: https://doi.org/10.1038/nclimate3096

[2] RITCHIE, Hannah: Cars, planes, trains: where do CO2 emissions from transport come from. In: *Our World in Data* (2020). https://ourworldindata.org/co2-emissions-from-transport. – Accessed on 7 July 2025.

[3] ERBACH, Gregor ; JENSEN, Liselotte: Fit for 55 package. In: *EPRS, European Parliament* (2022). https://vuphong.vn/wp-content/uploads/2023/09/Fit-for-553_EN.pdf. – Accessed on 7 July 2025.

[4] JUNG, Felix ; SCHRÖDER, Malte ; TIMME, Marc: Exponential adoption of battery electric cars. In: *PLoS one* 18 (2023), Nr. 12, S. e0295692. – DOI: https://doi.org/10.1371/journal.pone.0295692

[5] PESARAN, Ahmad A.: Battery thermal models for hybrid vehicle simulations. In: *Journal of Power Sources* 110 (2002), Nr. 2, S. 377–382. – DOI: https://doi.org/10.1016/S0378-7753(02)00200-8

[6] WALDMANN, Thomas ; WILKA, Marcel ; KASPER, Michael ; FLEISCHHAMMER, Meike ; WOHLFAHRT-MEHRENS, Margret: Temperature dependent ageing mechanisms in Lithium-ion batteries–A Post-Mortem study. In: *Journal of Power Sources* 262 (2014), S. 129–135. – DOI: https://doi.org/10.1016/j.jpowsour.2014.03.112

[7] LENG, Feng ; TAN, Cher M. ; PECHT, Michael: Effect of temperature on the aging rate of Li ion battery operating above room temperature. In: *Scientific Reports* 5 (2015), S. 12967. – DOI: https://doi.org/10.1038/srep12967

M. Auch, *Holistic Modeling Framework for the Investigation of Innovative Battery Thermal Management Concepts*, Wissenschaftliche Reihe Fahrzeugtechnik Universität Stuttgart, https://doi.org/10.1007/978-3-658-52612-2

[8] Qu, JG ; Jiang, ZY ; Zhang, JF: Investigation on lithium ion battery degradation induced by combined effect of current rate and operating temperature during fast charging. In: *Journal of Energy Storage* 52 (2022), S. 104811. – DOI: https://doi.org/10.1016/j.est.2022.104811

[9] Gharehghani, Ayat ; Rabiei, Moeed ; Mehranfar, Sadegh ; Saeedipour, Soheil ; Andwari, Amin M. ; García, Antonio ; Reche, Carlos M.: Progress in battery thermal management systems technologies for electric vehicles. In: *Renewable and Sustainable Energy Reviews* 202 (2024), S. 114654. – DOI: https://doi.org/10.1016/j.rser.2024.114654

[10] Wu, Chunxia ; Sun, Yalong ; Tang, Heng ; Zhang, Shiwei ; Yuan, Wei ; Zhu, Likuan ; Tang, Yong: A review on the liquid cooling thermal management system of lithium-ion batteries. In: *Applied Energy* 375 (2024), S. 124173. – DOI: https://doi.org/10.1016/j.apenergy.2024.124173

[11] Maiorino, Angelo ; Cilenti, Claudio ; Petruzziello, Fabio ; Aprea, Ciro: A review on thermal management of battery packs for electric vehicles. In: *Applied Thermal Engineering* 238 (2024), S. 122035. – DOI: https://doi.org/10.1016/j.applthermaleng.2023.122035

[12] Chang, Guofeng ; Cui, Xian ; Li, Yuyang ; Ji, Yunkang: Effects of reciprocating liquid flow battery thermal management system on thermal characteristics and uniformity of large lithium-ion battery pack. In: *International Journal of Energy Research* 44 (2020), Nr. 8, S. 6383–6395. – DOI: https://doi.org/10.1002/er.5363

[13] He, Yang ; Liu, Jianping ; Zhao, Xiaojian ; Zhang, Xiaoguang: *Battery Thermal Management System and Method.* https://patents.justia.com/patent/20220085438. Version: 2022. – Accessed on 18 November 2024.

[14] Liu, Wei ; Placke, Tobias ; Chau, KT: Overview of batteries and battery management for electric vehicles. In: *Energy Reports* 8 (2022), S. 4058–4084. – DOI: https://doi.org/10.1016/j.egyr.2022.03.016

[15] Wagner, Prof. Dr.-Ing. A. ; Krantz, Dr.-Ing. W.: *Grundlagen der Kraftfahrzeuge (Vorlesungsskript)*. 2023

[16] Hayes, John G. ; Goodarzi, G A.: Electric powertrain: energy systems, power electronics and drives for hybrid, electric and fuel cell vehicles. (2018). – ISBN: 9781119063643

[17] Wray, Alex ; Ebrahimi, Kambiz: Octovalve thermal management control for electric vehicle. In: *Energies* 15 (2022), Nr. 17, S. 6118. – DOI: https://doi.org/10.3390/en15176118

[18] Dr.-Ing. Christian Möser, Maximilian F.: Efficient Integral Thermal Management for Battery-Powered Electric Vehicles with Long Range. In: *Schaeffler Group* (2022). https://www.schaeffler.com/en/media/dates-events/kolloquium/digital-conference-book-2022/thermal-management-system-electric-vehicles. – Accessed on 2 May 2025.

[19] Wagner, Andreas: *Verbundvorhaben: CoolEV; Teilvorhaben: Gesamtsystemsimulation des externen Thermomanagements*. Institut für Fahrzeugtechnik, Universität Stuttgart, 2023. – DOI: https://doi.org/10.2314/KXP:1914654765

[20] Yoo, Hyun D. ; Markevich, Elena ; Salitra, Gregory ; Sharon, Daniel ; Aurbach, Doron: On the challenge of developing advanced technologies for electrochemical energy storage and conversion. In: *Materials today* 17 (2014), Nr. 3, S. 110–121. – DOI: https://doi.org/10.1016/j.mattod.2014.02.014

[21] JOUHARA, Hussam ; KHORDEHGAH, Navid ; SEREY, Nicolas ; ALMAHMOUD, Sulaiman ; LESTER, Stephen P. ; MACHEN, Daniel ; WROBEL, Luiz: Applications and thermal management of rechargeable batteries for industrial applications. In: *Energy* 170 (2019), S. 849–861. – DOI: https://doi.org/10.1016/j.energy.2018.12.218

[22] DORRMANN, Lydia ; SANN-FERRO, Kerstin ; HEININGER, Patrick ; MÄHLISS, Jochen: Kompendium: Li-ionen-batterien. In: *VDE Studie: Offenbach, Germany* (2021). https://www.dke.de/resource/blob/933404/fa7a24099c84ef613d8e7afd2c860a39/kompendium-li-ionen-batterien-data.pdf. – Accessed on 7 July 2025.

[23] WANG, Aiping ; KADAM, Sanket ; LI, Hong ; SHI, Siqi ; QI, Yue: Review on modeling of the anode solid electrolyte interphase (SEI) for lithium-ion batteries. In: *NPJ Computational materials* 4 (2018), Nr. 1, S. 15. – DOI: https://doi.org/10.1038/s41524-018-0064-0

[24] PLETT, Gregory L.: *Battery management systems, Volume I: Battery modeling*. Artech House, 2015. – ISBN: 9781630810245

[25] NUNES, Ashley ; SEE, Chung Y. ; WOODLEY, Lucas ; WANG, Seaver ; LIU, Gao: Estimating the tipping point for lithium iron phosphate batteries. In: *Applied Energy* 377 (2025), S. 124734. – DOI: https://doi.org/10.1016/j.apenergy.2024.124734

[26] MAJEED, Muhammad K. ; IQBAL, Rashid ; HUSSAIN, Arshad ; MAJEED, M U. ; ASHFAQ, M Z. ; AHMAD, Muhammad ; RAUF, Sajid ; SALEEM, Adil: Silicon-based anode materials for lithium batteries: recent progress, new trends, and future perspectives. In: *Critical Reviews in Solid State and Materials Sciences* 49 (2024), Nr. 2, S. 221–253. – DOI: https://doi.org/10.1080/10408436.2023.2169658

[27] BAAZOUZI, Sabri ; FEISTEL, Niklas ; WANNER, Johannes ; LANDWEHR, Inga ; FILL, Alexander ; BIRKE, Kai P.: Design, Properties, and Manufacturing of Cylindrical Li-Ion Battery CellsA Generic Overview. In: *Batteries* 9 (2023), Nr. 6, S. 309. – DOI: https://doi.org/10.3390/batteries9060309

[28] BMW Group: *Mehr Leistung, CO_2-Reduzierte Produktion, Kosten Deutlich Reduziert: Die BMW Group Setzt in der Neuen Klasse ab 2025 innovative BMW Batteriezellen im Rundformat ein.* https://www.press.bmwgroup.com/deutschland/article/detail/T0403470DE/mehr-leistung-co2-reduzierte-produktion-kosten-deutlich-reduziert-die-bmw-group-setzt-in-der-neuen. Version: 2022. – Accessed on 3 March 2025.

[29] Auch, Marcus ; Kuthada, Timo ; Giese, Sascha ; Wagner, Andreas: Influence of lithium-ion-battery equivalent circuit model parameter dependencies and architectures on the predicted heat generation in real-life drive cycles. In: *Batteries* 9 (2023), Nr. 5, S. 274. – DOI: https://doi.org/10.3390/batteries9050274

[30] Tran, Manh-Kien ; Mathew, Manoj ; Janhunen, Stefan ; Panchal, Satyam ; Raahemifar, Kaamran ; Fraser, Roydon ; Fowler, Michael: A comprehensive equivalent circuit model for lithium-ion batteries, incorporating the effects of state of health, state of charge, and temperature on model parameters. In: *Journal of Energy Storage* 43 (2021), S. 103252. – DOI: https://doi.org/10.1016/j.est.2021.103252

[31] Nejad, Shahab ; Gladwin, DT ; Stone, DA: A systematic review of lumped-parameter equivalent circuit models for real-time estimation of lithium-ion battery states. In: *Journal of Power Sources* 316 (2016), S. 183–196. – DOI: https://doi.org/10.1016/j.jpowsour.2016.03.042

[32] Mama, Magui ; Solai, Elie ; Capurso, Tommaso ; Danlos, Amelie ; Khelladi, Sofiane: Comprehensive review of multi-scale Lithium-ion batteries modeling: From electro-chemical dynamics up to heat transfer in battery thermal management system. In: *Energy Conversion and Management* 325 (2025), S. 119223. – DOI: https://doi.org/10.1016/j.enconman.2024.119223

[33] Lucaferri, Valentina ; Quercio, Michele ; Laudani, Antonino ; Riganti Fulginei, Francesco: A review on battery model-based and data-driven methods for battery management systems. In: *Energies* 16 (2023), Nr. 23, S. 7807. – DOI: https://doi.org/10.3390/en16237807

[34] Liu, Jie ; Yadav, Saurabh ; Salman, Mohammad ; Chavan, Santosh ; Kim, Sung C.: Review of thermal coupled battery models and parameter identification for lithium-ion battery heat generation in EV battery thermal management system. In: *International Journal of Heat and Mass Transfer* 218 (2024), S. 124748. – DOI: https://doi.org/10.1016/j.ijheatmasstransfer.2023.124748

[35] Zhang, Xiaoqiang ; Zhang, Weiping ; Lei, Geyang: A review of li-ion battery equivalent circuit models. In: *Transactions on Electrical and Electronic Materials* 17 (2016), Nr. 6, S. 311–316. – DOI: https://doi.org/10.4313/TEEM.2016.17.6.311

[36] Dong, TK ; Kirchev, A ; Mattera, F ; Kowal, Julia ; Bultel, Y: Dynamic modeling of Li-ion batteries using an equivalent electrical circuit. In: *Journal of the Electrochemical Society* 158 (2011), Nr. 3, S. A326. – DOI: https://doi.org/10.1149/1.3543710

[37] Reiter, Christoph: *Thermische Vorauslegung hochbelasteter Batteriesysteme für Elektrofahrzeuge in der Konzeptphase*, Technische Universität München, Diss., 2020. – ISBN: 9783844078381

[38] Hu, Xiaosong ; Li, Shengbo ; Peng, Huei: A comparative study of equivalent circuit models for Li-ion batteries. In: *Journal of Power Sources* 198 (2012), S. 359–367. – DOI: https://doi.org/10.1016/j.jpowsour.2011.10.013

[39] Kassem, M ; Delacourt, C: Postmortem analysis of calendar-aged graphite/LiFePO4 cells. In: *Journal of Power Sources* 235 (2013), S. 159–171. – DOI: https://doi.org/10.1016/j.jpowsour.2013.01.147

[40] VETTER, Jens ; NOVÁK, Petr ; WAGNER, Markus R. ; VEIT, Claudia ; MÖLLER, K-C ; BESENHARD, JO ; WINTER, Martin ; WOHLFAHRT-MEHRENS, Margret ; VOGLER, Christoph ; HAMMOUCHE, Abderrezak: Ageing mechanisms in lithium-ion batteries. In: *Journal of Power Sources* 147 (2005), Nr. 1-2, S. 269–281. – DOI: `https://doi.org/10.1016/j.jpowsour.2005.01.006`

[41] BÜLOW, Friedrich von: *A Data-Driven Fleet Service: State of Health Forecasting of Lithium-Ion Batteries.* Springer, 2024. – DOI: `https://doi.org/10.1007/978-3-658-43188-4`

[42] KUCINSKIS, Gints ; BOZORGCHENANI, Maral ; FEINAUER, Max ; KASPER, Michael ; WOHLFAHRT-MEHRENS, Margret ; WALDMANN, Thomas: Arrhenius plots for Li-ion battery ageing as a function of temperature, C-rate, and ageing state–An experimental study. In: *Journal of Power Sources* 549 (2022), S. 232129. – DOI: `https://doi.org/10.1016/j.jpowsour.2022.232129`

[43] ECKER, Madeleine ; GERSCHLER, Jochen B. ; VOGEL, Jan ; KÄBITZ, Stefan ; HUST, Friedrich ; DECHENT, Philipp ; SAUER, Dirk U.: Development of a lifetime prediction model for lithium-ion batteries based on extended accelerated aging test data. In: *Journal of Power Sources* 215 (2012), S. 248–257. – DOI: `https://doi.org/10.1016/j.jpowsour.2012.05.012`

[44] BIRKL, Christoph R. ; ROBERTS, Matthew R. ; MCTURK, Euan ; BRUCE, Peter G. ; HOWEY, David A.: Degradation diagnostics for lithium ion cells. In: *Journal of Power Sources* 341 (2017), S. 373–386. – DOI: `https://doi.org/10.1016/j.jpowsour.2016.12.011`

[45] FLECKENSTEIN, Matthias ; BOHLEN, Oliver ; BÄKER, Bernard: Aging effect of temperature gradients in Li-ion cells experimental and simulative investigations and the consequences on thermal battery management. In: *World Electric Vehicle Journal* 5 (2012), Nr. 2, S. 322–333. – DOI: `https://doi.org/10.3390/wevj5020322`

[46] Bernardi, D ; Pawlikowski, E ; Newman, John: A general energy balance for battery systems. In: *Journal of the electrochemical society* 132 (1985), Nr. 1, S. 5. – DOI: https://doi.org/10.1149/1.2113792

[47] Spotnitz, R ; Franklin, J: Abuse behavior of high-power, lithium-ion cells. In: *Journal of power sources* 113 (2003), Nr. 1, S. 81–100. – DOI: https://doi.org/10.1016/S0378-7753(02)00488-3

[48] Thomas, Karen E. ; Newman, John: Thermal modeling of porous insertion electrodes. In: *Journal of the Electrochemical Society* 150 (2003), Nr. 2, S. A176. – DOI: https://doi.org/10.1149/1.1531194

[49] Lagnoni, Marco ; Scarpelli, Claudio ; Lutzemberger, Giovanni ; Bertei, Antonio: Critical comparison of equivalent circuit and physics-based models for lithium-ion batteries: A graphite/lithium-iron-phosphate case study. In: *Journal of Energy Storage* 94 (2024), S. 112326. – DOI: https://doi.org/10.1016/j.est.2024.112326

[50] Ank, Manuel ; Sommer, Alessandro ; Gamra, Kareem A. ; Schöberl, Jan ; Leeb, Matthias ; Schachtl, Johannes ; Streidel, Noah ; Stock, Sandro ; Schreiber, Markus ; Bilfinger, Philip u. a.: Lithium-ion cells in automotive applications: Tesla 4680 cylindrical cell teardown and characterization. In: *Journal of The Electrochemical Society* 170 (2023), Nr. 12, S. 120536. – DOI: https://doi.org/10.1149/1945-7111/ad14d0

[51] Steinhardt, Marco ; Barreras, Jorge V. ; Ruan, Haijun ; Wu, Billy ; Offer, Gregory J. ; Jossen, Andreas: Meta-analysis of experimental results for heat capacity and thermal conductivity in lithium-ion batteries: A critical review. In: *Journal of Power Sources* 522 (2022), S. 230829. – DOI: https://doi.org/10.1016/j.jpowsour.2021.230829

[52] He, Tengfei ; Zhang, Teng ; Wang, Zhirong ; Cai, Qiong: A comprehensive numerical study on electrochemical-thermal models of a cylindrical lithium-ion battery during discharge process. In: *Applied Energy* 313 (2022), S. 118797. – DOI: https://doi.org/10.1016/j.apenergy.2022.118797

[53] Alkhedher, Mohammad ; Al Tahhan, Aghyad B. ; Yousaf, Jawad ; Ghazal, Mohammed ; Shahbazian-Yassar, Reza ; Ramadan, Mohamad: Electrochemical and thermal modeling of lithium-ion batteries: A review of coupled approaches for improved thermal performance and safety lithium-ion batteries. In: *Journal of Energy Storage* 86 (2024), S. 111172. – DOI: https://doi.org/10.1016/j.est.2024.111172

[54] Yang, Xiaolong ; Hu, Xiaobo ; Chen, Zhuo ; Chen, Yongji: Effect of ambient dissipation condition on thermal behavior of a lithium-ion battery using a 3D multi-partition model. In: *Applied Thermal Engineering* 178 (2020), S. 115634. – DOI: https://doi.org/10.1016/j.applthermaleng.2020.115634

[55] Dubey, Dwijendra ; Mishra, A ; Ghosh, Subrata ; Reddy, MV ; Pandey, Ramesh: Geometry-influenced cooling performance of lithium-ion battery. In: *Applied Thermal Engineering* 230 (2023), S. 120723. – DOI: https://doi.org/10.1016/j.applthermaleng.2023.120723

[56] Khaboshan, Hasan N. ; Jaliliantabar, Farzad ; Abdullah, Abdul A. ; Panchal, Satyam: Improving the cooling performance of cylindrical lithium-ion battery using three passive methods in a battery thermal management system. In: *Applied Thermal Engineering* 227 (2023), S. 120320. – DOI: https://doi.org/10.1016/j.applthermaleng.2023.120320

[57] Zeng, Wei ; Niu, Yi ; Li, Silin ; Hu, Sihang ; Mao, Binbin ; Zhang, Ying: Cooling performance and optimization of a new hybrid thermal management system of cylindrical battery. In: *Applied Thermal Engineering* 217 (2022), S. 119171. – DOI: https://doi.org/10.1016/j.applthermaleng.2022.119171

[58] XIN, Shiji ; WANG, Chun ; XI, Huan: Thermal management scheme and optimization of cylindrical lithium-ion battery pack based on air cooling and liquid cooling. In: *Applied Thermal Engineering* 224 (2023), S. 120100. – DOI: https://doi.org/10.1016/j.applthermaleng.2023.120100

[59] LI, Li ; LING, Lei ; XIE, Yajun ; ZHOU, Wencai ; WANG, Tianbo ; ZHANG, Lanchun ; BEI, Shaoyi ; ZHENG, Keqing ; XU, Qiang: Comparative study of thermal management systems with different cooling structures for cylindrical battery modules: Side-cooling vs. terminal-cooling. In: *Energy* 274 (2023), S. 127414. – DOI: https://doi.org/10.1016/j.energy.2023.127414

[60] ROBINSON, James B. ; DARR, Jawwad A. ; EASTWOOD, David S. ; HINDS, Gareth ; LEE, Peter D. ; SHEARING, Paul R. ; TAIWO, Oluwadamilola O. ; BRETT, Dan J.: Non-uniform temperature distribution in Li-ion batteries during discharge–A combined thermal imaging, X-ray micro-tomography and electrochemical impedance approach. In: *Journal of Power Sources* 252 (2014), S. 51–57. – DOI: https://doi.org/10.1016/j.jpowsour.2013.11.059

[61] NAYLOR MARLOW, Max ; CHEN, Jingyi ; WU, Billy: Degradation in parallel-connected lithium-ion battery packs under thermal gradients. In: *Communications Engineering* 3 (2024), Nr. 1, S. 2. – DOI: https://doi.org/10.1038/s44172-023-00153-5

[62] WU, Weixiong ; WANG, Shuangfeng ; WU, Wei ; CHEN, Kai ; HONG, Sihui ; LAI, Yongxin: A critical review of battery thermal performance and liquid based battery thermal management. In: *Energy conversion and management* 182 (2019), S. 262–281. – DOI: https://doi.org/10.1016/j.enconman.2018.12.051

[63] KHAN, Aamir ; YAQUB, Sana ; ALI, Majid ; AHMAD, Adeel W. ; NAZIR, Hassan ; KHALID, Hassan A. ; IQBAL, Naseem ; SAID, Zafar ; SOPIAN, Kamaruzzaman: A state-of-the-art review on heating and cooling of lithium-ion batteries for electric vehicles. In: *Journal of Energy Storage* 76 (2024), S. 109852. – DOI: https://doi.org/10.1016/j.est.2023.109852

[64] Subramanian, Mohankumar ; Ağbulut, Ümit ; Pachamuthu, Senthilkumar ; Sathanandam, Satish ; Solomon, Jenoris M. ; Arputharaj, Beena S. ; Raja, Vijayanandh ; Rajendran, Parvathy ; Asif, Mohammad: Computational and experimental investigations on liquid-based battery thermal management systems for electric vehicle applications under various discharge rates with different flow speeds. In: *Journal of Energy Storage* 91 (2024), S. 111757. – DOI: `https://doi.org/10.1016/j.est.2024.111757`

[65] Hannoschöck, Nikolaus: *Wärmeleitung und-transport: Grundlagen der Wärme-und Stoffübertragung*. Springer-Verlag, 2018. – ISBN: 9783662575710

[66] Stephan, Peter: B2 Grundlagen der Berechnungsmethoden für Wärmeleitung, konvektiven Wärmeübergang und Wärmestrahlung. In: *VDI-Wärmeatlas: Fachlicher Träger VDI-Gesellschaft Verfahrenstechnik und Chemieingenieurwesen* (2019), S. 23–36. – DOI: `https://doi.org/10.1007/978-3-662-52989-8_4`

[67] Kalogirou, Soteris A.: *Solar energy engineering: processes and systems*. Academic press, 2013. – DOI: `https://doi.org/10.1016/B978-0-12-374501-9.X0001-5`

[68] Lee, Poh-Seng ; Garimella, Suresh V.: Thermally developing flow and heat transfer in rectangular microchannels of different aspect ratios. In: *International journal of heat and mass transfer* 49 (2006), Nr. 17-18, S. 3060–3067. – DOI: `https://doi.org/10.1016/j.ijheatmasstransfer.2006.02.011`

[69] Verein Deutscher Ingenieure, VDI-Gesellschaft Verfahrenstechnik und Chemieingenieurwesen (. ; Klan, H: Wärmeübergang durch freie Konvektion an umströmten Körpern. In: *VDI-Wärmeatlas: Berechnungsblätter für den Wärmeübergang* (2002), S. 567–591. – DOI: `https://doi.org/10.1007/978-3-662-10743-0_6`

[70] Rennels, Donald C.: *Pipe flow: A practical and comprehensive guide*. John Wiley & Sons, 2022. – DOI: `https://doi.org/10.1002/9781118275276`

[71] AUCH, Marcus ; WEYERSHÄUSER, Konstantin ; KUTHADA, Timo ; WAGNER, Andreas: Reducing Temperature-related Aging Inhomogeneities in Battery Modules Using a Switchable Thermal Management System / SAE Technical Paper. 2025. – Forschungsbericht. – DOI: https://doi.org/10.4271/2025-01-8167

[72] OMARIBA, Zachary B. ; ZHANG, Lijun ; SUN, Dongbai: Review of battery cell balancing methodologies for optimizing battery pack performance in electric vehicles. In: *Ieee Access* 7 (2019), S. 129335–129352. – DOI: https://doi.org/10.1109/ACCESS.2019.2940090

[73] ZHUANG, Weichao ; LIU, Zhitao ; SU, Hongye ; CHEN, Guangwei: An intelligent thermal management system for optimized lithium-ion battery pack. In: *Applied Thermal Engineering* 189 (2021), S. 116767. – DOI: https://doi.org/10.1016/j.applthermaleng.2021.116767

[74] ZENG, Wei ; MA, Chuyuan ; HU, Sihang ; LI, Silin ; ZHANG, Ying: The performance investigation and optimization of reciprocating flow applied for liquid-cooling-based battery thermal management system. In: *Energy Conversion and Management* 292 (2023), S. 117378. – DOI: https://doi.org/10.1016/j.enconman.2023.117378

[75] CHEN, Saihan ; SUN, Jinlei ; QIU, Shengshi ; LIU, Xinwei ; LYU, Kai ; CHEN, Siwen ; XING, Shiyou ; GUO, Yilong: Phased control reciprocating airflow cooling strategy for a battery module considering stage of charge and state of health inconsistency. In: *Journal of Energy Storage* 61 (2023), S. 106752. – DOI: https://doi.org/10.1016/j.est.2023.106752

[76] HE, Haosong ; FLY, Ashley ; BARBOUR, Edward ; CHEN, Xiangjie: Numerical investigation of module-level inhomogeneous ageing in lithium-ion batteries from temperature gradients and electrical connection topologies. In: *Communications Engineering* 3 (2024), Nr. 1, S. 89. – DOI: https://doi.org/10.1038/s44172-024-00222-3

[77] Jinasena, Asanthi ; Spitthoff, Lena ; Wahl, Markus S. ; Lamb, Jacob J. ; Shearing, Paul R. ; Strømman, Anders H. ; Burheim, Odne S.: Online internal temperature sensors in lithium-ion batteries: State-of-the-art and future trends. In: *Frontiers in Chemical Engineering* 4 (2022), S. 804704. – DOI: https://doi.org/10.3389/fceng.2022.804704

[78] Auch, Marcus ; Kuthada, Timo ; Wagner, Andreas: Simple experimental method to determine the specific heat capacity of cylindrical Lithium-Ion-Battery cells. In: *Applied Thermal Engineering* (2023), S. 121212. – DOI: https://doi.org/10.1016/j.applthermaleng.2023.121212

[79] Auch, Marcus ; Kuthada, Timo ; Wagner, Andreas: Improved equivalent circuit coupled 3D thermal cylindrical lithium-ion battery model using a multi-partition heat generation and thermal resistance approach. In: *Applied Thermal Engineering* 247 (2024), S. 123056. – DOI: https://doi.org/10.1016/j.applthermaleng.2024.123056

[80] Talele, Virendra ; Ahmadian-Elmi, Mohammad ; Jain, Ankur ; Mishra, Dhananjay ; Zawodzinski Jr, Thomas A. ; Zhao, Peng: Simulation and reduction of nonuniform capacity fading in a cylindrical battery module via reciprocating thermal design. In: *Journal of Energy Storage* 103 (2024), S. 114143. – DOI: https://doi.org/10.1016/j.est.2024.114143

[81] EV Database: *Tesla Model 3 Long Range Dual Motor*. https://ev-database.org/imp/car/1591/Tesla-Model-3-Long-Range-Dual-Motor, 2024. – Accessed on 14 May 2025.

[82] Ha, Jong H.: PRODUCT SPECIFICATION Rechargeable Lithium Ion Battery Model : INR21700 M50T 18.2Wh. (2018). https://www.batteryspace.com/prod-specs/11514.pdf. – Accessed on 18/10/2022.

[83] Arcus, Christopher: *Tesla Model 3 & Chevy Bolt Battery Packs Examined.* https://cleantechnica.com/2018/07/08/tesla-model-3-chevy-bolt-battery-packs-examined/. Version: 2018. – Accessed on 20 April 2023

[84] Wassiliadis, Nikolaos ; Ank, Manuel ; Wildfeuer, Leo ; Kick, Michael K. ; Lienkamp, Markus: Experimental investigation of the influence of electrical contact resistance on lithium-ion battery testing for fast-charge applications. In: *Applied Energy* 295 (2021), S. 117064. – DOI: https://doi.org/10.1016/j.apenergy.2021.117064

[85] Doughty, Daniel H. ; Crafts, Chris C.: FreedomCAR electrical energy storage system abuse test manual for electric and hybrid electric vehicle applications / Sandia National Laboratories. 2006. – Forschungsbericht. – DOI: https://doi.org/10.2172/889934

[86] Barai, Anup ; Uddin, Kotub ; Widanage, Widanalage D. ; McGordon, Andrew ; Jennings, Paul: A study of the influence of measurement timescale on internal resistance characterisation methodologies for lithium-ion cells. In: *Scientific reports* 8 (2018), Nr. 1, S. 21. – DOI: https://doi.org/10.1038/s41598-017-18424-5

[87] He, Hongwen ; Xiong, Rui ; Fan, Jinxin: Evaluation of lithium-ion battery equivalent circuit models for state of charge estimation by an experimental approach. In: *energies* 4 (2011), Nr. 4, S. 582–598. – DOI: https://doi.org/10.3390/en4040582

[88] Drake, SJ ; Wetz, DA ; Ostanek, JK ; Miller, SP ; Heinzel, JM ; Jain, A: Measurement of anisotropic thermophysical properties of cylindrical Li-ion cells. In: *Journal of Power Sources* 252 (2014), S. 298–304. – DOI: https://doi.org/10.1016/j.jpowsour.2013.11.107

[89] Steinhardt, Marco ; Gillich, Elisabeth I. ; Rheinfeld, Alexander ; Kraft, Ludwig ; Spielbauer, Markus ; Bohlen, Oliver ; Jossen, Andreas: Low-effort determination of heat capacity and thermal conductivity for cylindrical 18650 and 21700 lithium-ion cells. In: *Journal of Energy Storage* 42 (2021), S. 103065. – DOI: https://doi.org/10.1016/j.est.2021.103065

[90] TOOLBOX, The E.: Metal Alloys - Specific Heats. (2003). https://www.engineeringtoolbox.com/specific-heat-metal-alloys-d_153.html

[91] TOOLBOX, The E.: Metals, Metallic Elements and Alloys - Thermal Conductivities. (2005). https://www.engineeringtoolbox.com/thermal-conductivity-metals-d_858.html

[92] WANG, John ; LIU, Ping ; HICKS-GARNER, Jocelyn ; SHERMAN, Elena ; SOUKIAZIAN, Souren ; VERBRUGGE, Mark ; TATARIA, Harshad ; MUSSER, James ; FINAMORE, Peter: Cycle-life model for graphite-LiFePO4 cells. In: *Journal of Power Sources* 196 (2011), Nr. 8, S. 3942–3948. – DOI: https://doi.org/10.1016/j.jpowsour.2010.11.134

[93] LUH, Matthias ; BLANK, Thomas: Comprehensive battery aging dataset: capacity and impedance fade measurements of a lithium-ion NMC/C-SiO cell. In: *Scientific Data* 11 (2024), S. 1004. – DOI: https://doi.org/10.1038/s41597-024-03831-x

[94] PAARMANN, Sabine ; SCHULD, Kathrin ; WETZEL, Thomas: Inhomogeneous Aging in Lithium-Ion Batteries Caused by Temperature Effects. In: *Energy Technology* 10 (2022), Nr. 11, S. 2200384. – DOI: https://doi.org/10.1002/ente.202200384

[95] PEGEL, Hendrik ; AUTENRIETH, Moritz ; SCHAEFFLER, Stefan ; JOSSEN, Andreas ; SAUER, Dirk U.: Design guidelines to prevent thermal propagation and maximize packing density within battery systems with tabless cylindrical lithium-ion cells. In: *Journal of Energy Storage* 86 (2024), S. 111275. – DOI: https://doi.org/10.1016/j.est.2024.111275

[96] YANG, Bruce ; KARIMI, Hamid ; BAGHERI, Kamran: *Integrated Energy Storage System*. https://uspto.report/patent/app/20210159567. Version: 2021. – Patent Application

[97] SIEMENS, PLM: STAR-CCM+ User Guide Version 13.04. In: *Siemens PLM Software Inc: Munich, Germany* (2022)

[98] SIEMENS INDUSTRY SOFTWARE NV: *User's Guide of AMESim Rev12*. 2022

[99] Wassiliadis, Nikolaos ; Steinsträter, Matthias ; Schreiber, Markus ; Rosner, Philipp ; Nicoletti, Lorenzo ; Schmid, Florian ; Ank, Manuel ; Teichert, Olaf ; Wildfeuer, Leo ; Schneider, Jakob u. a.: Quantifying the state of the art of electric powertrains in battery electric vehicles: Range, efficiency, and lifetime from component to system level of the Volkswagen ID. 3. In: *Etransportation* 12 (2022), S. 100167. – DOI: https://doi.org/10.1016/j.etran.2022.100167

[100] Europe, Economic C.: Proposal for a new global technical regulation on the Worldwide harmonized Light vehicles Test Procedure (WLTP). In: *World Forum for Harmonization of Vehicle Regulations 162nd session* (2014)

[101] Huang, Jian ; Xu, Peichen ; Wang, Peiyong: Experimental measurement of anisotropic thermal conductivity of 18650 lithium battery. In: *Journal of Physics: Conference Series* Bd. 1509 IOP Publishing, 2020, S. 012013. – DOI: https://doi.org/10.1088/1742-6596/1509/1/012013

[102] Zhu, Linpei ; Xiong, Fei ; Chen, Hu ; Wei, Dan ; Li, Gang ; Ouyang, Chenzhi: Thermal analysis and optimization of an EV battery pack for real applications. In: *International Journal of Heat and Mass Transfer* 163 (2020), S. 120384. – DOI: https://doi.org/10.1016/j.ijheatmasstransfer.2020.120384

[103] Barcellona, S ; Piegari, L: Effect of current on cycle aging of lithium ion batteries. In: *Journal of Energy Storage* 29 (2020), S. 101310. – DOI: https://doi.org/10.1016/j.est.2020.101310

[104] Redondo-Iglesias, Eduardo ; Venet, Pascal ; Pelissier, Serge: Calendar and cycling ageing combination of batteries in electric vehicles. In: *Microelectronics Reliability* 88 (2018), S. 1212–1215. – DOI: https://doi.org/10.1016/j.microrel.2018.06.113

[105] Kirkaldy, Niall ; Samieian, Mohammad A. ; Offer, Gregory J. ; Marinescu, Monica ; Patel, Yatish: Lithium-ion battery degradation: Comprehensive cycle ageing data and analysis for commercial 21700 cells. In: *Journal of Power Sources* 603 (2024), S. 234185. – DOI: https://doi.org/10.1016/j.jpowsour.2024.234185

[106] Frenander, Kristian ; Thiringer, Torbjörn: Analysis of the number of replicates required for Li-ion battery degradation testing. In: *Journal of Energy Storage* 102 (2024), S. 114014. – DOI: https://doi.org/10.1016/j.est.2024.114014

[107] Baumhöfer, Thorsten ; Brühl, Manuel ; Rothgang, Susanne ; Sauer, Dirk U.: Production caused variation in capacity aging trend and correlation to initial cell performance. In: *Journal of Power Sources* 247 (2014), S. 332–338. – DOI: https://doi.org/10.1016/j.jpowsour.2013.08.108

[108] Schreiber, Markus ; Lin, Yilei ; Sommer, Alessandro ; Wassiliadis, Nikolaos ; Torricos, Pablo M. ; Rogge, Marcel ; Lewerenz, Meinert ; Grosu, Cristina ; Endisch, Christian ; Jossen, Andreas u. a.: Apparent vs. true battery aging: Impact of various load characteristics on accelerated aging tests. In: *Journal of Energy Storage* 127 (2025), S. 116924. – DOI: https://doi.org/10.1016/j.est.2025.116924

[109] Iurilli, Pietro: *EIS applied to Li-Ion Batteries: Modelling Cell Aging for Comprehensive SoH Estimation*, ETH Zurich, Diss., 2022. – DOI: https://doi.org/10.3929/ethz-b-000589948

[110] Tendera, Luca ; , Hendrik ; Gonzalez, Carlos ; Wycisk, Dominik ; Fill, Alexander ; Birke, Kai P.: Influence of temperature, state of charge and state of health on the thermal parameters of lithium-ion cells: Exploring thermal behavior and enabling fast-charging. In: *Future Batteries* 1 (2024), S. 100001. – DOI: https://doi.org/10.1016/j.fub.2024.100001

[111] Kakaç, Sadık ; Shah, Ramesh K. ; Aung, Win: Handbook of single-phase convective heat transfer. (1987). – ISBN: 9780471817024

[112] Bennett, Ted D.: Correlations for the Graetz problem in convection–Part 2: For ducts of arbitrary cross-section. In: *International Journal of Heat and Mass Transfer* 135 (2019), S. 1327–1334. – doi: https://doi.org/10.1016/j.ijheatmasstransfer.2019.02.052

[113] Khoshvaght-Aliabadi, M ; Sahamiyan, M ; Hesampour, M ; Sartipzadeh, O: Experimental study on cooling performance of sinusoidal–wavy minichannel heat sink. In: *Applied Thermal Engineering* 92 (2016), S. 50–61. – DOI: https://doi.org/10.1016/j.applthermaleng.2015.09.015

[114] Sui, Y ; Teo, CJ ; Lee, Poh S. ; Chew, YT ; Shu, C: Fluid flow and heat transfer in wavy microchannels. In: *International Journal of Heat and Mass Transfer* 53 (2010), Nr. 13-14, S. 2760–2772. – DOI: https://doi.org/10.1016/j.ijheatmasstransfer.2010.02.022

[115] Love, Corey T. ; Virji, Maheboob B. ; Rocheleau, Richard E. ; Swider-Lyons, Karen E.: State-of-health monitoring of 18650 4S packs with a single-point impedance diagnostic. In: *Journal of Power Sources* 266 (2014), S. 512–519. – DOI: https://doi.org/10.1016/j.jpowsour.2014.05.033

[116] George Bower, Inside EV: Tesla Model S Plaid Battery: Clever New Advancements Discovered. (2022). https://insideevs.com/news/566047/tesla-models-clever-battery-advancements/. – Accessed on 08 September 2025.

[117] Hadała, Beata ; Zygmunt-Kowalska, Beata ; Kuźnia, Monika ; Szajding, Artur ; Telejko, Tadeusz: Thermal insulation properties of rigid polyurethane foam modified with fly ash-a comparative study. In: *Thermochimica Acta* 731 (2024), S. 179659. – DOI: https://doi.org/10.1016/j.tca.2023.179659

[118] The Engineering ToolBox: Ethylene Glycol Heat-Transfer Fluid Properties. (2003). https://www.engineeringtoolbox.com/ethylene-glycol-d_146.html. – Accessed on 12 September 2024.

Appendix

A. HPPC

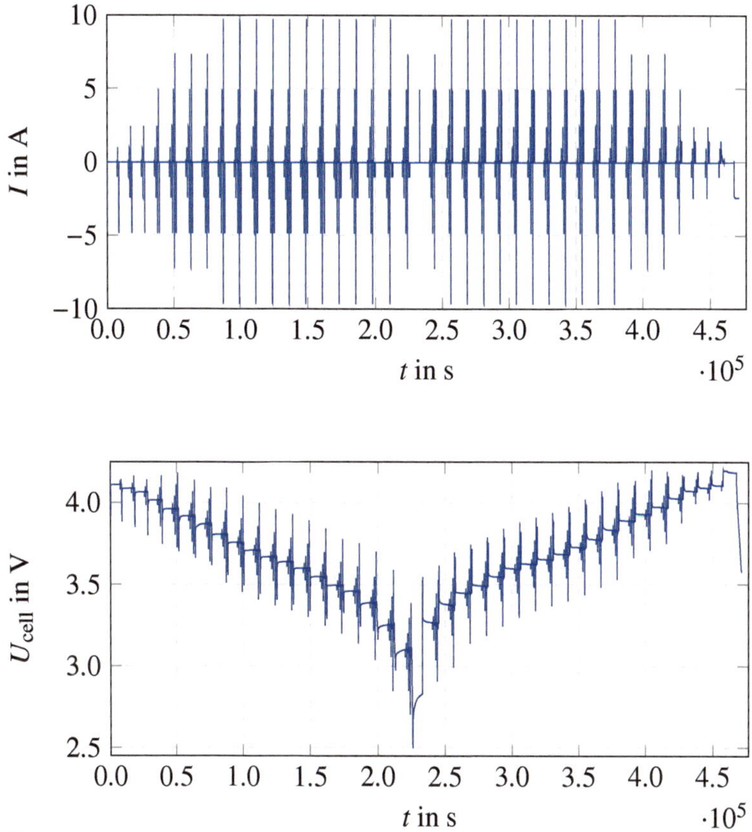

Figure A.1: Hybrid Pulse Power Characterization (HPPC) results for current and voltage using 5% SOC-steps for discharge and charge at $T_{\mathrm{amb}} = 25\,°\mathrm{C}$.

M. Auch, *Holistic Modeling Framework for the Investigation of Innovative Battery Thermal Management Concepts*, Wissenschaftliche Reihe Fahrzeugtechnik Universität Stuttgart, https://doi.org/10.1007/978-3-658-52612-2

B. Single cell ECM

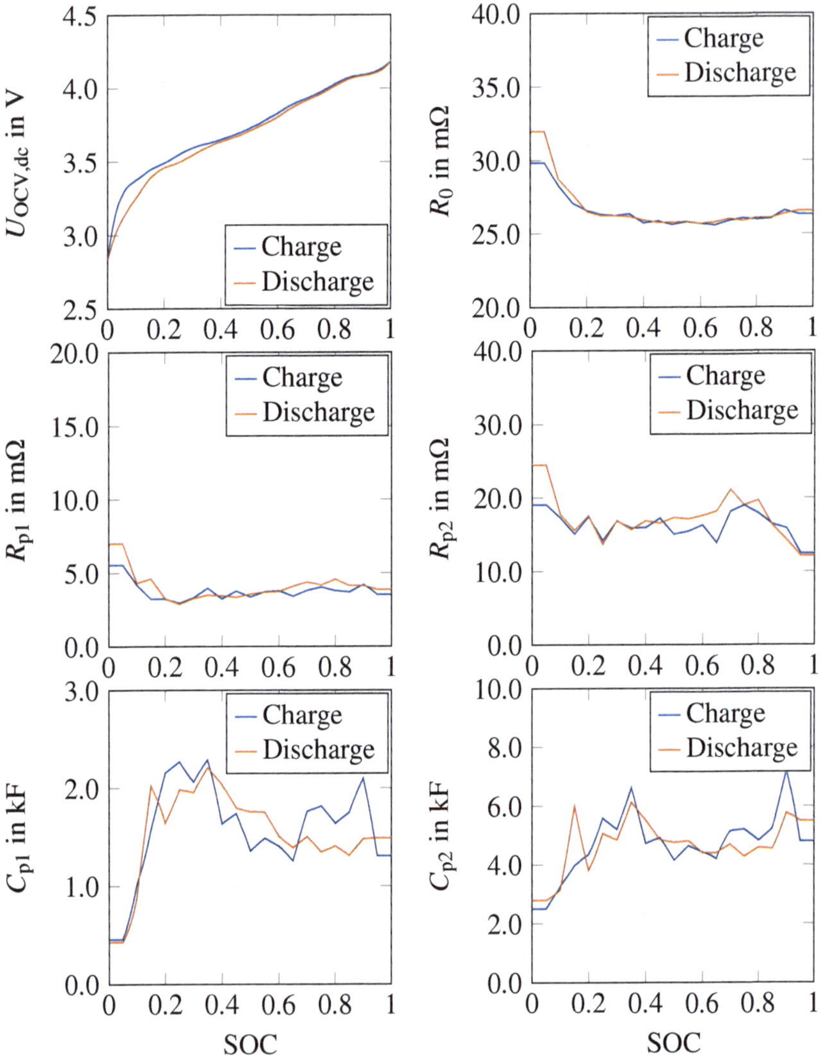

Figure B.1: Derived ECM parameters for the single cell during charge and discharge at $T = 25\,°C$, determined using a 0.5 C pulse.

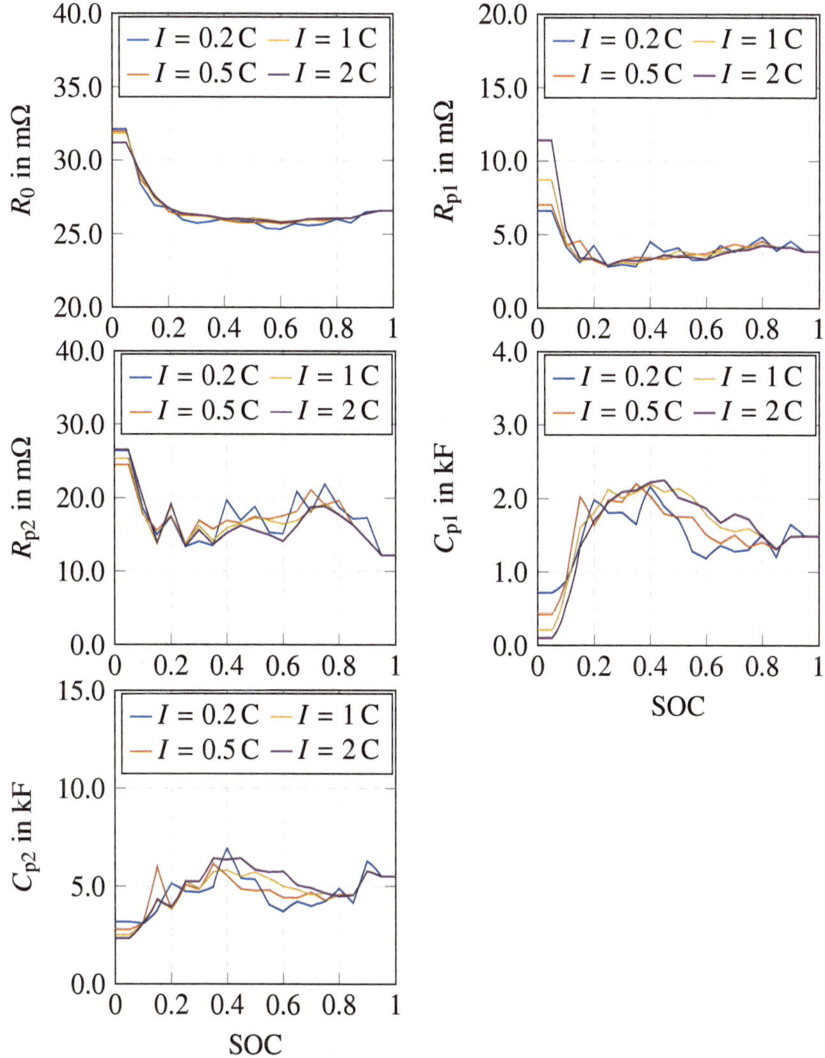

Figure B.2: Derived ECM parameters for the single cell for various discharge current pulses $I = 0.2, 0.5, 1, 2\,\mathrm{C}$ at $T = 25\,°\mathrm{C}$.

Table B.1: Measured capacity for the single cell at various temperatures.

Temperature T in °C	5	25	45
Capacity Q in Ah	4.59	4.80	4.72

Table B.2: Derived ECM OCV parameters for the single cell during charge and discharge at various temperatures and SOC steps.

SOC	T in °C	$U_{\mathrm{OCV,dc}}$ in V	$U_{\mathrm{OCV,ch}}$ in V
100	[5,25,45]	[4.186,4.185,4.181]	[4.180,4.186,4.188]
95	[5,25,45]	[4.109,4.111,4.110]	[4.117,4.119,4.115]
90	[5,25,45]	[4.086,4.088,4.088]	[4.092,4.094,4.093]
85	[5,25,45]	[4.069,4.065,4.067]	[4.077,4.075,4.074]
80	[5,25,45]	[4.026,4.017,4.020]	[4.039,4.031,4.030]
75	[5,25,45]	[3.973,3.963,3.967]	[3.987,3.978,3.978]
70	[5,25,45]	[3.929,3.920,3.923]	[3.942,3.933,3.934]
65	[5,25,45]	[3.889,3.872,3.877]	[3.904,3.893,3.893]
60	[5,25,45]	[3.825,3.808,3.816]	[3.851,3.836,3.836]
55	[5,25,45]	[3.772,3.759,3.768]	[3.792,3.779,3.785]
50	[5,25,45]	[3.723,3.712,3.720]	[3.743,3.731,3.736]
45	[5,25,45]	[3.681,3.672,3.679]	[3.698,3.688,3.693]
40	[5,25,45]	[3.648,3.641,3.643]	[3.663,3.656,3.660]
35	[5,25,45]	[3.622,3.602,3.598]	[3.637,3.628,3.630]
30	[5,25,45]	[3.586,3.551,3.552]	[3.612,3.599,3.595]
25	[5,25,45]	[3.528,3.498,3.499]	[3.574,3.550,3.548]
20	[5,25,45]	[3.481,3.463,3.465]	[3.524,3.495,3.494]
15	[5,25,45]	[3.444,3.389,3.393]	[3.469,3.448,3.448]
10	[5,25,45]	[3.339,3.254,3.265]	[3.428,3.378,3.371]
5	[5,25,45]	[3.197,3.103,3.114]	[3.329,3.268,3.264]
0	[5,25,45]	[3.042,2.833,2.814]	[3.042,2.833,2.814]

Table B.3: Derived discharge ECM parameters for the single cell at various temperatures, determined using various current pulses.

SOC	T in °C	I in A	R_0 in mΩ	R_{p1} in mΩ	R_{p2} in mΩ	C_{p1} in kF	C_{p2} in kF
100	[5,25,45]	0.96	[39.01,26.63,24.30]	[6.43,3.89,2.48]	[22.52,12.23,6.66]	[0.33,1.50,1.04]	[2.88,5.53,4.87]
	[5,25,45]	2.4	[39.01,26.63,24.30]	[6.43,3.89,2.48]	[22.52,12.23,6.66]	[0.33,1.50,1.04]	[2.88,5.53,4.87]
	[5,25,45]	4.8	[39.01,26.63,24.30]	[6.43,3.89,2.48]	[22.52,12.23,6.66]	[0.33,1.50,1.04]	[2.88,5.53,4.87]
	[5,25,45]	9.6	[39.01,26.63,24.30]	[6.43,3.89,2.48]	[22.52,12.23,6.66]	[0.33,1.50,1.04]	[2.88,5.53,4.87]
95	[5,25,45]	0.96	[39.01,26.63,24.30]	[6.43,3.89,2.48]	[22.52,12.23,6.66]	[0.33,1.50,1.04]	[2.88,5.53,4.87]
	[5,25,45]	2.4	[39.01,26.63,24.30]	[6.43,3.89,2.48]	[22.52,12.23,6.66]	[0.33,1.50,1.04]	[2.88,5.53,4.87]
	[5,25,45]	4.8	[39.01,26.63,24.30]	[6.43,3.89,2.48]	[22.52,12.23,6.66]	[0.33,1.50,1.04]	[2.88,5.53,4.87]
	[5,25,45]	9.6	[39.01,26.63,24.30]	[6.43,3.89,2.48]	[22.52,12.23,6.66]	[0.33,1.50,1.04]	[2.88,5.53,4.87]
90	[5,25,45]	0.96	[37.49,26.52,24.34]	[5.22,4.62,2.86]	[20.87,17.38,8.88]	[0.68,1.66,1.49]	[3.22,6.31,6.24]
	[5,25,45]	2.4	[37.49,26.41,24.74]	[5.22,4.19,3.56]	[20.87,14.38,9.78]	[0.68,1.49,1.93]	[3.22,5.79,9.03]
	[5,25,45]	4.8	[37.49,26.41,24.74]	[5.22,4.19,3.56]	[20.87,14.38,9.78]	[0.68,1.49,1.93]	[3.22,5.79,9.03]
	[5,25,45]	9.6	[37.49,26.41,24.74]	[5.22,4.19,3.56]	[20.87,14.38,9.78]	[0.68,1.49,1.93]	[3.22,5.79,9.03]
85	[5,25,45]	0.96	[36.32,25.81,24.65]	[5.51,3.96,4.60]	[27.08,17.27,16.02]	[0.95,1.21,1.92]	[3.34,4.18,9.28]
	[5,25,45]	2.4	[36.62,26.12,24.22]	[5.91,4.19,2.88]	[28.29,16.37,9.48]	[0.97,1.32,1.27]	[3.48,4.58,4.74]
	[5,25,45]	4.8	[36.62,26.12,24.22]	[5.91,4.19,2.88]	[28.29,16.37,9.48]	[0.97,1.32,1.27]	[3.48,4.58,4.74]
	[5,25,45]	9.6	[36.62,26.12,24.22]	[5.91,4.19,2.88]	[28.29,16.37,9.48]	[0.97,1.32,1.27]	[3.48,4.58,4.74]
80	[5,25,45]	0.96	[36.12,26.06,24.40]	[6.22,4.89,3.59]	[31.95,18.84,12.46]	[1.16,1.52,1.62]	[3.50,4.91,5.97]
	[5,25,45]	2.4	[35.94,26.13,24.21]	[5.97,4.61,3.00]	[30.91,19.75,10.91]	[1.05,1.41,1.28]	[3.32,4.61,4.54]
	[5,25,45]	4.8	[35.94,26.10,24.47]	[5.97,4.30,3.35]	[30.91,17.80,11.27]	[1.05,1.49,1.69]	[3.32,4.51,5.56]
	[5,25,45]	9.6	[35.94,26.10,24.47]	[5.97,4.30,3.35]	[30.91,17.80,11.27]	[1.05,1.49,1.69]	[3.32,4.51,5.56]
75	[5,25,45]	0.96	[35.73,25.70,24.70]	[6.04,4.31,4.64]	[33.38,21.95,15.49]	[1.19,1.31,2.10]	[3.38,4.26,7.79]
	[5,25,45]	2.4	[35.59,25.95,24.35]	[5.82,4.22,3.32]	[31.59,19.14,12.84]	[1.06,1.35,1.57]	[3.29,4.31,5.31]
	[5,25,45]	4.8	[35.59,26.12,24.44]	[5.82,4.34,3.20]	[31.59,19.41,12.22]	[1.06,1.61,1.72]	[3.29,4.68,5.41]
	[5,25,45]	9.6	[35.59,26.07,24.48]	[5.82,4.03,3.27]	[31.59,18.92,12.19]	[1.06,1.73,2.04]	[3.29,4.68,5.96]
70	[5,25,45]	0.96	[35.29,25.61,24.06]	[5.73,3.89,3.02]	[34.26,18.23,12.08]	[1.16,1.29,1.40]	[3.36,4.02,5.11]
	[5,25,45]	2.4	[35.56,26.06,24.43]	[6.09,4.42,3.29]	[33.69,21.18,12.54]	[1.23,1.51,1.67]	[3.62,4.71,5.60]
	[5,25,45]	4.8	[35.39,25.97,24.48]	[5.60,3.98,3.29]	[31.71,18.37,12.86]	[1.22,1.57,1.88]	[3.55,4.60,5.96]
	[5,25,45]	9.6	[35.39,26.06,24.52]	[5.60,3.95,3.10]	[31.71,18.81,13.10]	[1.22,1.80,2.40]	[3.55,4.94,6.52]
65	[5,25,45]	0.96	[35.01,25.75,24.12]	[5.44,4.32,3.15]	[32.71,20.96,11.13]	[1.08,1.37,1.25]	[3.15,4.25,4.56]
	[5,25,45]	2.4	[35.34,25.83,24.23]	[5.79,4.11,3.19]	[31.71,18.26,11.40]	[1.18,1.40,1.37]	[3.49,4.44,4.85]
	[5,25,45]	4.8	[35.22,25.94,24.23]	[5.40,3.93,2.84]	[29.83,17.02,10.42]	[1.19,1.62,1.51]	[3.53,4.87,4.91]
	[5,25,45]	9.6	[34.83,25.88,24.31]	[5.26,3.72,2.74]	[28.57,16.24,10.37]	[1.10,1.68,2.01]	[3.60,5.08,5.90]
60	[5,25,45]	0.96	[35.01,25.38,24.16]	[4.88,3.32,2.92]	[31.38,15.21,11.14]	[1.09,1.19,1.68]	[3.13,3.74,5.08]
	[5,25,45]	2.4	[35.04,25.74,24.33]	[4.90,3.76,3.29]	[28.38,17.66,12.65]	[1.15,1.51,1.93]	[3.35,4.44,5.93]
	[5,25,45]	4.8	[35.10,25.87,24.27]	[4.82,3.62,2.71]	[27.74,16.55,10.43]	[1.29,1.79,2.01]	[3.75,5.03,6.01]
	[5,25,45]	9.6	[34.67,25.82,24.29]	[4.68,3.32,2.57]	[25.65,14.17,8.96]	[1.19,1.88,2.17]	[3.97,5.77,7.29]
55	[5,25,45]	0.96	[35.19,25.43,23.92]	[4.85,3.30,2.50]	[27.44,15.38,8.77]	[1.32,1.29,1.31]	[3.77,4.08,4.43]
	[5,25,45]	2.4	[35.14,25.85,24.27]	[4.58,3.73,2.99]	[25.38,17.22,11.57]	[1.18,1.76,1.84]	[3.61,4.84,5.58]
	[5,25,45]	4.8	[35.10,26.03,24.37]	[4.58,3.80,2.92]	[24.58,17.00,10.85]	[1.27,2.03,2.20]	[3.84,5.43,6.13]
	[5,25,45]	9.6	[34.71,25.93,24.35]	[4.56,3.50,2.67]	[23.10,15.02,10.00]	[1.14,1.96,2.18]	[3.95,5.74,6.45]

50	[5,25,45]	0.96	[34.92,25.88,23.85]	[4.13,4.16,2.58]	[23.11,18.90,8.97]	[1.14,1.73,1.34]	[3.36,5.38,4.70]
	[5,25,45]	2.4	[35.23,25.82,24.22]	[4.64,3.60,2.71]	[27.20,17.37,11.34]	[1.29,1.76,1.70]	[3.76,4.80,5.17]
	[5,25,45]	4.8	[35.10,26.13,24.65]	[4.37,3.96,3.48]	[24.84,17.65,12.21]	[1.26,2.15,2.53]	[3.75,5.75,7.40]
	[5,25,45]	9.6	[34.48,26.00,24.46]	[4.58,3.53,2.76]	[24.08,15.65,11.18]	[1.16,2.02,2.26]	[4.08,5.86,6.67]
45	[5,25,45]	0.96	[35.10,25.95,24.61]	[4.50,3.89,4.00]	[26.77,17.03,12.18]	[1.29,1.90,2.35]	[3.89,5.45,9.03]
	[5,25,45]	2.4	[35.08,25.82,24.33]	[4.27,3.37,2.78]	[25.68,16.65,11.08]	[1.19,1.80,1.92]	[3.68,4.88,5.83]
	[5,25,45]	4.8	[35.32,25.98,24.60]	[4.52,3.52,3.18]	[26.51,16.59,12.09]	[1.42,2.10,2.52]	[4.08,5.54,7.27]
	[5,25,45]	9.6	[34.56,26.07,24.64]	[4.55,3.66,3.06]	[24.81,16.30,11.79]	[1.20,2.26,2.60]	[4.26,6.45,8.09]
40	[5,25,45]	0.96	[35.14,26.08,24.48]	[4.29,4.57,3.33]	[24.86,19.79,10.36]	[1.33,2.17,2.22]	[4.06,6.96,8.32]
	[5,25,45]	2.4	[35.21,25.96,24.28]	[4.20,3.49,2.38]	[24.88,16.98,9.32]	[1.23,2.04,1.76]	[3.89,5.55,5.47]
	[5,25,45]	4.8	[35.32,26.03,24.52]	[4.35,3.42,2.64]	[25.39,16.05,9.85]	[1.43,2.21,2.36]	[4.23,5.86,6.90]
	[5,25,45]	9.6	[34.66,26.05,24.46]	[4.48,3.33,2.41]	[24.01,15.29,8.58]	[1.22,2.23,2.16]	[4.46,6.37,7.13]
35	[5,25,45]	0.96	[34.94,25.88,24.32]	[3.75,2.88,2.58]	[20.33,13.60,9.38]	[0.97,1.66,1.85]	[3.61,4.99,5.71]
	[5,25,45]	2.4	[35.49,26.21,24.60]	[4.32,3.53,3.07]	[25.05,15.80,10.75]	[1.39,2.21,2.28]	[4.22,6.16,7.03]
	[5,25,45]	4.8	[35.40,26.15,24.67]	[4.12,3.06,3.00]	[23.09,14.28,10.44]	[1.33,2.10,2.39]	[4.20,5.78,7.29]
	[5,25,45]	9.6	[34.79,26.24,24.56]	[4.42,3.26,2.70]	[22.53,13.71,9.58]	[1.15,2.12,2.06]	[4.52,6.44,6.57]
30	[5,25,45]	0.96	[35.39,25.77,24.23]	[3.91,3.00,2.33]	[21.41,14.14,10.56]	[1.14,1.82,1.47]	[3.99,4.71,4.13]
	[5,25,45]	2.4	[35.64,26.27,24.41]	[4.06,3.30,2.56]	[22.06,17.01,10.91]	[1.24,1.96,1.76]	[4.10,4.88,4.74]
	[5,25,45]	4.8	[35.54,26.33,24.55]	[4.06,3.20,2.61]	[21.48,16.29,11.08]	[1.21,2.02,1.97]	[4.16,4.91,5.10]
	[5,25,45]	9.6	[34.81,26.31,24.58]	[4.28,3.28,2.60]	[20.68,15.68,10.45]	[0.94,2.10,2.03]	[4.14,5.26,5.53]
25	[5,25,45]	0.96	[35.88,25.98,24.59]	[4.09,2.85,3.14]	[23.48,13.44,9.82]	[1.33,1.82,2.16]	[4.06,4.76,6.71]
	[5,25,45]	2.4	[35.80,26.29,24.48]	[4.06,2.92,2.36]	[24.13,13.78,9.23]	[1.09,1.99,1.80]	[3.76,5.10,5.18]
	[5,25,45]	4.8	[36.02,26.43,24.62]	[4.23,2.90,2.35]	[25.10,13.88,9.33]	[1.26,2.14,2.08]	[4.02,5.31,5.59]
	[5,25,45]	9.6	[35.18,26.41,24.66]	[4.35,2.94,2.40]	[23.69,13.44,9.02]	[0.87,1.95,1.94]	[3.94,5.26,5.69]
20	[5,25,45]	0.96	[36.44,26.81,25.11]	[3.75,4.30,4.58]	[24.50,17.52,10.89]	[1.01,1.99,2.18]	[3.39,5.17,7.72]
	[5,25,45]	2.4	[36.51,26.53,24.65]	[4.07,3.28,2.69]	[25.71,17.64,10.64]	[1.02,1.65,1.47]	[3.52,3.85,4.09]
	[5,25,45]	4.8	[36.40,26.74,24.80]	[4.19,3.42,2.69]	[25.58,19.07,11.62]	[0.98,1.82,1.65]	[3.59,3.89,4.09]
	[5,25,45]	9.6	[35.25,26.76,24.89]	[4.80,3.38,2.67]	[24.37,19.28,12.07]	[0.49,1.71,1.73]	[3.33,3.97,4.38]
15	[5,25,45]	0.96	[38.96,26.98,25.29]	[4.80,3.13,4.42]	[30.56,15.06,11.68]	[1.39,1.37,2.00]	[3.68,3.78,8.95]
	[5,25,45]	2.4	[37.90,27.68,25.01]	[4.90,4.63,2.62]	[31.37,15.70,9.00]	[0.61,2.03,1.51]	[3.02,6.02,4.85]
	[5,25,45]	4.8	[36.74,27.45,25.26]	[6.00,3.30,2.81]	[29.81,13.87,8.61]	[0.26,1.62,1.87]	[2.83,4.41,5.74]
	[5,25,45]	9.6	[34.92,27.57,25.46]	[7.85,3.43,2.99]	[27.75,13.97,8.69]	[0.08,1.35,1.80]	[2.75,4.35,6.07]
10	[5,25,45]	0.96	[36.55,28.46,25.18]	[11.05,4.23,3.08]	[26.61,18.67,10.09]	[0.05,0.88,0.76]	[2.30,3.13,3.38]
	[5,25,45]	2.4	[34.40,28.75,26.49]	[14.83,4.34,5.44]	[26.62,17.85,10.14]	[0.02,0.88,1.84]	[2.19,3.18,9.10]
	[5,25,45]	4.8	[34.40,29.07,25.93]	[14.83,4.53,3.32]	[26.62,18.19,10.64]	[0.02,0.80,1.19]	[2.19,3.16,4.19]
	[5,25,45]	9.6	[34.40,29.24,26.99]	[14.83,5.33,5.50]	[26.62,20.09,11.60]	[0.02,0.56,1.82]	[2.19,3.14,8.66]
5	[5,25,45]	0.96	[40.64,32.19,27.70]	[27.51,6.66,6.07]	[47.10,26.50,16.37]	[0.03,0.73,1.65]	[1.63,3.22,8.73]
	[5,25,45]	2.4	[40.64,32.02,27.89]	[27.51,7.07,5.91]	[47.10,24.57,13.99]	[0.03,0.44,1.69]	[1.63,2.82,8.55]
	[5,25,45]	4.8	[40.64,31.90,28.14]	[27.51,8.77,5.95]	[47.10,25.47,14.54]	[0.03,0.22,1.68]	[1.63,2.57,8.45]
	[5,25,45]	9.6	[40.64,31.23,28.57]	[27.51,11.45,6.36]	[47.10,26.58,21.49]	[0.03,0.11,1.57]	[1.63,2.37,7.10]
0	[5,25,45]	0.96	[40.64,32.19,27.70]	[27.51,6.66,6.07]	[47.10,26.50,16.37]	[0.03,0.73,1.65]	[1.63,3.22,8.73]
	[5,25,45]	2.4	[40.64,32.02,27.89]	[27.51,7.07,5.91]	[47.10,24.57,13.99]	[0.03,0.44,1.69]	[1.63,2.82,8.55]
	[5,25,45]	4.8	[40.64,31.90,28.14]	[27.51,8.77,5.95]	[47.10,25.47,14.54]	[0.03,0.22,1.68]	[1.63,2.57,8.45]
	[5,25,45]	9.6	[40.64,31.23,28.57]	[27.51,11.45,6.36]	[47.10,26.58,21.49]	[0.03,0.11,1.57]	[1.63,2.37,7.10]

Table B.4: Derived charge ECM parameters for the single cell at various temperatures, determined using various current pulses.

SOC	T in °C	I in A	R_0 in mΩ	R_{p1} in mΩ	R_{p2} in mΩ	C_{p1} in kF	C_{p2} in kF
100	[5,25,45]	0.96	[40.16,26.36,24.80]	[6.08,3.54,3.40]	[25.46,12.55,9.30]	[0.53,1.31,1.92]	[3.21,4.82,8.25]
	[5,25,45]	2.4	[40.16,26.36,24.80]	[6.08,3.54,3.40]	[25.46,12.55,9.30]	[0.53,1.31,1.92]	[3.21,4.82,8.25]
	[5,25,45]	4.8	[40.16,26.36,24.80]	[6.08,3.54,3.40]	[25.46,12.55,9.30]	[0.53,1.31,1.92]	[3.21,4.82,8.25]
	[5,25,45]	9.6	[40.16,26.36,24.80]	[6.08,3.54,3.40]	[25.46,12.55,9.30]	[0.53,1.31,1.92]	[3.21,4.82,8.25]
95	[5,25,45]	0.96	[40.16,26.36,24.80]	[6.08,3.54,3.40]	[25.46,12.55,9.30]	[0.53,1.31,1.92]	[3.21,4.82,8.25]
	[5,25,45]	2.4	[40.16,26.36,24.80]	[6.08,3.54,3.40]	[25.46,12.55,9.30]	[0.53,1.31,1.92]	[3.21,4.82,8.25]
	[5,25,45]	4.8	[40.16,26.36,24.80]	[6.08,3.54,3.40]	[25.46,12.55,9.30]	[0.53,1.31,1.92]	[3.21,4.82,8.25]
	[5,25,45]	9.6	[40.16,26.36,24.80]	[6.08,3.54,3.40]	[25.46,12.55,9.30]	[0.53,1.31,1.92]	[3.21,4.82,8.25]
90	[5,25,45]	0.96	[37.86,26.75,24.45]	[5.41,4.92,2.73]	[24.17,18.06,9.29]	[0.86,1.91,1.72]	[3.71,7.29,6.43]
	[5,25,45]	2.4	[37.86,26.62,24.49]	[5.41,4.25,2.76]	[24.17,15.96,9.32]	[0.86,2.10,2.21]	[3.71,7.24,7.90]
	[5,25,45]	4.8	[37.86,26.62,24.49]	[5.41,4.25,2.76]	[24.17,15.96,9.32]	[0.86,2.10,2.21]	[3.71,7.24,7.90]
	[5,25,45]	9.6	[37.86,26.62,24.49]	[5.41,4.25,2.76]	[24.17,15.96,9.32]	[0.86,2.10,2.21]	[3.71,7.24,7.90]
85	[5,25,45]	0.96	[36.84,26.41,24.17]	[5.60,4.80,2.89]	[29.46,20.27,10.15]	[1.13,1.74,1.39]	[3.75,5.84,5.46]
	[5,25,45]	2.4	[36.46,26.09,24.27]	[5.26,3.72,2.70]	[27.25,16.51,10.07]	[1.04,1.76,1.96]	[3.78,5.27,6.12]
	[5,25,45]	4.8	[36.46,26.09,24.27]	[5.26,3.72,2.70]	[27.25,16.51,10.07]	[1.04,1.76,1.96]	[3.78,5.27,6.12]
	[5,25,45]	9.6	[36.46,26.09,24.27]	[5.26,3.72,2.70]	[27.25,16.51,10.07]	[1.04,1.76,1.96]	[3.78,5.27,6.12]
80	[5,25,45]	0.96	[35.97,26.29,24.70]	[5.67,4.83,4.02]	[32.89,20.96,14.02]	[1.14,1.71,2.06]	[3.45,5.25,7.35]
	[5,25,45]	2.4	[35.81,26.02,24.70]	[5.58,3.81,3.88]	[29.91,17.94,12.99]	[1.17,1.65,2.37]	[3.75,4.85,8.15]
	[5,25,45]	4.8	[35.81,25.96,24.37]	[5.58,3.79,3.10]	[29.91,16.86,11.23]	[1.17,1.80,2.29]	[3.75,5.97,8.31]
	[5,25,45]	9.6	[35.81,25.96,24.37]	[5.58,3.79,3.10]	[29.91,16.86,11.23]	[1.17,1.80,2.29]	[3.75,5.97,8.31]
75	[5,25,45]	0.96	[35.58,25.94,24.68]	[5.26,3.83,4.73]	[31.20,18.31,17.75]	[1.07,1.35,2.12]	[3.22,4.30,8.44]
	[5,25,45]	2.4	[35.31,26.10,24.35]	[5.20,4.04,3.05]	[28.90,19.05,12.52]	[1.08,1.82,2.02]	[3.49,5.23,6.20]
	[5,25,45]	4.8	[35.31,25.85,24.39]	[5.20,3.65,3.18]	[28.90,16.58,12.28]	[1.08,1.77,2.39]	[3.49,5.85,8.55]
	[5,25,45]	9.6	[35.31,25.89,24.48]	[5.20,3.79,3.23]	[28.90,14.84,10.17]	[1.08,1.92,2.56]	[3.49,7.02,10.91]
70	[5,25,45]	0.96	[35.25,25.72,24.25]	[5.13,3.72,3.34]	[28.71,18.48,14.36]	[1.12,1.36,1.78]	[3.38,4.33,6.20]
	[5,25,45]	2.4	[35.00,25.93,24.36]	[5.06,3.82,3.09]	[28.16,18.24,11.91]	[1.14,1.77,2.11]	[3.63,5.17,6.70]
	[5,25,45]	4.8	[34.21,25.90,24.37]	[5.02,3.77,3.08]	[26.13,17.35,11.61]	[0.95,1.98,2.51]	[3.88,6.33,8.75]
	[5,25,45]	9.6	[34.21,25.83,24.30]	[5.02,3.61,2.69]	[26.13,14.61,7.80]	[0.95,1.97,2.39]	[3.88,7.07,10.34]
65	[5,25,45]	0.96	[35.09,25.94,24.54]	[4.87,4.12,3.99]	[24.90,16.28,13.82]	[1.03,1.43,1.83]	[3.27,5.28,7.34]
	[5,25,45]	2.4	[34.93,25.61,24.24]	[5.04,3.44,3.05]	[27.42,13.95,10.85]	[1.15,1.26,1.59]	[3.56,4.23,5.41]
	[5,25,45]	4.8	[34.06,25.55,24.05]	[4.85,3.25,2.60]	[25.94,14.94,10.33]	[0.92,1.49,1.64]	[3.65,4.54,5.25]
	[5,25,45]	9.6	[33.55,25.46,24.04]	[4.79,3.02,2.29]	[24.25,14.71,9.34]	[0.87,1.59,2.00]	[3.91,4.91,6.43]
60	[5,25,45]	0.96	[35.60,26.07,24.39]	[6.08,4.88,3.65]	[33.50,20.04,13.04]	[1.37,1.67,1.60]	[3.95,5.04,5.72]
	[5,25,45]	2.4	[34.86,25.74,24.33]	[4.83,3.82,3.22]	[26.58,16.37,11.04]	[1.08,1.41,1.66]	[3.47,4.46,5.84]
	[5,25,45]	4.8	[33.96,25.59,24.14]	[4.72,3.49,2.86]	[25.41,14.87,9.68]	[0.85,1.45,1.58]	[3.44,4.72,6.04]
	[5,25,45]	9.6	[33.42,25.41,24.08]	[4.69,2.98,2.46]	[25.34,13.74,8.22]	[0.79,1.62,1.80]	[3.53,4.97,6.21]
55	[5,25,45]	0.96	[35.17,25.88,24.15]	[4.87,4.08,3.00]	[28.82,17.41,11.78]	[1.18,1.42,1.35]	[3.41,4.54,4.51]
	[5,25,45]	2.4	[35.21,25.84,24.66]	[5.00,3.75,4.00]	[27.62,15.52,12.48]	[1.31,1.50,2.12]	[3.76,4.66,7.08]
	[5,25,45]	4.8	[34.16,25.65,24.22]	[4.72,3.46,2.83]	[24.57,14.55,10.26]	[0.88,1.49,1.71]	[3.59,4.87,6.00]
	[5,25,45]	9.6	[33.55,25.47,24.11]	[4.74,3.28,2.49]	[23.53,13.15,8.10]	[0.78,1.49,1.56]	[3.66,5.32,6.33]

50	[5,25,45]	0.96	[35.51,25.81,24.77]	[5.25,3.72,4.82]	[32.02,16.31,19.86]	[1.43,1.47,2.05]	[3.69,4.11,7.55]
	[5,25,45]	2.4	[35.27,25.67,24.45]	[4.94,3.39,3.29]	[29.36,15.13,12.42]	[1.38,1.36,1.81]	[3.75,4.18,5.81]
	[5,25,45]	4.8	[34.23,25.75,24.25]	[4.64,3.51,2.80]	[26.20,14.75,10.85]	[0.96,1.59,1.72]	[3.62,5.04,5.81]
	[5,25,45]	9.6	[33.24,25.54,24.26]	[4.45,3.24,2.52]	[22.48,13.09,8.73]	[0.84,1.47,1.73]	[3.77,5.34,6.71]
45	[5,25,45]	0.96	[35.81,25.88,24.26]	[5.65,3.99,2.84]	[31.79,21.65,11.12]	[1.65,1.64,1.45]	[4.12,4.55,4.91]
	[5,25,45]	2.4	[35.06,25.90,24.44]	[4.49,3.79,3.00]	[27.74,17.33,11.08]	[1.26,1.75,1.73]	[3.62,4.95,5.79]
	[5,25,45]	4.8	[34.22,25.68,24.36]	[4.47,3.35,2.83]	[25.89,14.71,10.81]	[0.92,1.63,1.87]	[3.62,5.06,6.31]
	[5,25,45]	9.6	[33.39,25.52,24.44]	[4.43,3.08,2.58]	[23.58,12.75,9.09]	[0.91,1.50,2.08]	[3.98,5.41,7.71]
40	[5,25,45]	0.96	[35.73,25.78,24.30]	[5.79,3.65,2.72]	[33.84,17.41,10.34]	[1.68,1.73,1.45]	[4.59,4.72,5.06]
	[5,25,45]	2.4	[35.10,25.77,24.50]	[4.27,3.27,2.97]	[25.67,16.06,10.33]	[1.24,1.64,1.92]	[3.75,4.75,6.45]
	[5,25,45]	4.8	[34.33,25.79,24.29]	[4.37,3.32,2.65]	[25.35,14.89,9.47]	[0.98,1.82,1.79]	[3.85,5.54,6.36]
	[5,25,45]	9.6	[33.50,25.63,24.32]	[4.28,3.02,2.32]	[22.97,12.87,8.74]	[0.93,1.74,1.97]	[4.14,5.96,6.84]
35	[5,25,45]	0.96	[35.09,26.07,24.66]	[4.02,3.44,2.78]	[23.59,17.08,10.72]	[1.13,1.74,2.26]	[3.65,5.56,6.09]
	[5,25,45]	2.4	[35.30,26.38,24.30]	[4.38,3.98,2.44]	[25.40,16.01,10.26]	[1.40,2.29,1.73]	[4.15,6.64,5.48]
	[5,25,45]	4.8	[34.49,26.01,24.47]	[4.31,3.22,2.93]	[24.20,14.69,10.96]	[1.00,1.99,2.25]	[4.03,5.72,7.17]
	[5,25,45]	9.6	[33.46,25.73,24.44]	[4.15,2.89,2.58]	[21.54,13.43,9.28]	[0.84,1.86,2.30]	[4.14,5.77,7.41]
30	[5,25,45]	0.96	[35.51,25.87,24.43]	[3.88,3.06,2.83]	[23.10,17.40,11.37]	[1.14,1.73,1.90]	[3.83,4.71,5.50]
	[5,25,45]	2.4	[35.36,26.27,24.44]	[4.02,3.34,2.63]	[22.03,16.90,11.24]	[1.22,2.07,1.94]	[4.11,5.23,5.15]
	[5,25,45]	4.8	[34.44,26.03,24.24]	[4.21,3.15,2.42]	[21.78,15.82,10.32]	[0.85,1.92,1.59]	[4.03,5.09,4.74]
	[5,25,45]	9.6	[33.46,25.89,24.34]	[4.09,2.94,2.42]	[19.94,14.79,9.99]	[0.71,1.87,1.88]	[3.97,5.20,5.33]
25	[5,25,45]	0.96	[35.99,26.21,24.60]	[3.92,2.65,2.46]	[25.07,13.70,10.55]	[1.33,1.82,2.21]	[3.91,5.06,5.70]
	[5,25,45]	2.4	[35.59,26.33,24.43]	[3.95,2.98,2.14]	[23.66,14.24,9.22]	[1.17,2.28,1.87]	[3.88,5.62,5.16]
	[5,25,45]	4.8	[34.57,26.10,24.32]	[4.28,2.69,2.16]	[23.44,13.40,8.85]	[0.73,1.98,1.76]	[3.70,5.30,5.36]
	[5,25,45]	9.6	[33.71,25.94,24.33]	[4.21,2.69,2.12]	[22.41,12.56,8.13]	[0.68,1.89,1.81]	[3.75,5.50,5.65]
20	[5,25,45]	0.96	[36.19,26.43,24.54]	[3.66,3.09,2.60]	[22.63,16.26,11.58]	[0.89,1.79,1.56]	[3.46,3.87,4.10]
	[5,25,45]	2.4	[35.96,26.60,24.59]	[3.90,3.27,2.48]	[23.30,17.51,10.62]	[0.94,2.16,1.74]	[3.71,4.38,4.25]
	[5,25,45]	4.8	[34.42,26.25,24.52]	[4.55,2.72,2.35]	[21.95,15.51,10.18]	[0.43,1.86,1.87]	[3.52,4.29,4.65]
	[5,25,45]	9.6	[33.44,26.19,24.53]	[4.55,2.71,2.18]	[21.03,13.98,9.03]	[0.41,1.91,1.96]	[3.63,4.86,5.23]
15	[5,25,45]	0.96	[38.05,27.35,24.83]	[4.12,3.63,2.42]	[27.54,15.14,9.26]	[0.92,1.81,1.21]	[3.15,4.32,3.66]
	[5,25,45]	2.4	[35.68,27.11,24.78]	[5.45,3.26,2.48]	[25.91,15.16,8.98]	[0.23,1.58,1.32]	[2.86,4.01,4.05]
	[5,25,45]	4.8	[33.72,26.71,24.72]	[6.65,3.09,2.56]	[25.91,14.07,8.74]	[0.11,1.35,1.38]	[2.95,3.88,4.48]
	[5,25,45]	9.6	[33.25,26.45,24.59]	[6.41,2.90,2.41]	[23.96,13.19,8.18]	[0.10,1.39,1.34]	[3.06,3.89,4.44]
10	[5,25,45]	0.96	[36.90,27.98,25.23]	[9.41,3.86,2.94]	[28.22,15.02,8.72]	[0.07,0.68,0.81]	[2.35,2.81,3.32]
	[5,25,45]	2.4	[35.96,28.28,25.28]	[9.28,4.21,3.11]	[26.09,17.36,10.01]	[0.07,1.02,0.91]	[2.33,3.29,3.60]
	[5,25,45]	4.8	[35.96,27.65,25.07]	[9.28,4.16,3.10]	[26.09,17.10,9.64]	[0.07,0.84,0.84]	[2.33,3.07,3.47]
	[5,25,45]	9.6	[35.96,27.10,24.99]	[9.28,3.80,3.06]	[26.09,16.15,9.69]	[0.07,0.85,0.89]	[2.33,2.90,3.47]
5	[5,25,45]	0.96	[38.80,29.85,25.71]	[18.06,5.84,3.40]	[31.71,17.89,9.16]	[0.04,0.33,0.40]	[1.46,2.33,2.85]
	[5,25,45]	2.4	[38.80,29.88,25.98]	[18.06,5.60,3.46]	[31.71,19.15,10.41]	[0.04,0.46,0.58]	[1.46,2.53,3.24]
	[5,25,45]	4.8	[38.80,28.84,25.66]	[18.06,5.67,3.55]	[31.71,19.15,10.36]	[0.04,0.42,0.54]	[1.46,2.51,3.19]
	[5,25,45]	9.6	[38.80,28.37,25.57]	[18.06,5.38,3.49]	[31.71,19.04,10.56]	[0.04,0.48,0.66]	[1.46,2.55,3.31]
0	[5,25,45]	0.96	[38.80,29.85,25.71]	[18.06,5.84,3.40]	[31.71,17.89,9.16]	[0.04,0.33,0.40]	[1.46,2.33,2.85]
	[5,25,45]	2.4	[38.80,29.88,25.98]	[18.06,5.60,3.46]	[31.71,19.15,10.41]	[0.04,0.46,0.58]	[1.46,2.53,3.24]
	[5,25,45]	4.8	[38.80,28.84,25.66]	[18.06,5.67,3.55]	[31.71,19.15,10.36]	[0.04,0.42,0.54]	[1.46,2.51,3.19]
	[5,25,45]	9.6	[38.80,28.37,25.57]	[18.06,5.38,3.49]	[31.71,19.04,10.56]	[0.04,0.48,0.66]	[1.46,2.55,3.31]

Table B.5: Derived entropic coefficient parameters for the single cell at various SOC steps.

U_{OCV} in V	SOC in %	$\partial U_{\mathrm{OCV}}/\partial T$ in mV/K
4.178	100	0.090
4.107	95	0.109
4.087	90	0.090
4.065	85	0.065
4.015	80	0.070
3.962	75	0.080
3.918	70	0.070
3.861	65	0.159
3.804	60	0.199
3.755	55	0.199
3.707	50	0.199
3.667	45	0.169
3.635	40	0.035
3.590	35	-0.055
3.536	30	-0.080
3.488	25	-0.055
3.442	20	-0.298
3.351	15	-0.383
3.226	10	-0.438
3.046	5	-0.522
2.827	0	-0.602

C. ECM cell-to-cell

Table C.1: Cell-to-cell variation and arithmetic mean of measured capacity at $T = 25\,°C$.

Cell	2	3	4	5	6	Mean
Capacity Q in Ah	4.76	4.75	4.72	4.73	4.73	4.74

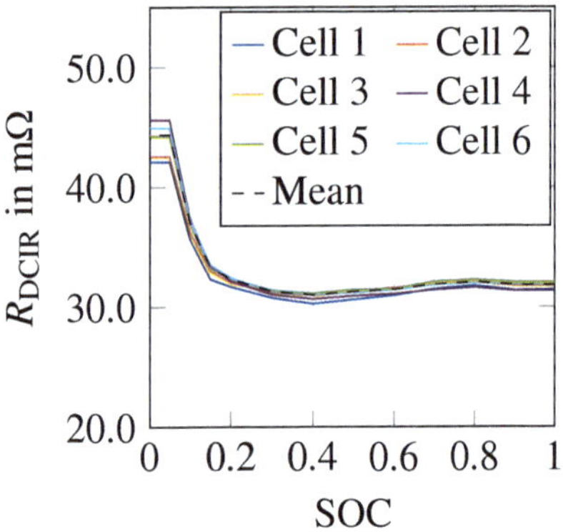

Figure C.1: Cell-to-cell variation of DC internal resistance R_{DCIR} based on the Rint model, determined using a 0.5 C discharge pulse.

Table C.2: Mean average error (MAE) and mean relative error (MRE) of the DC internal resistance R_{DCIR} of all six cells, compared to the arithmetic mean value, for a 0.5 C discharge pulse at $T = 25\,°C$ using the Rint ECM.

	Cell 1	Cell 2	Cell 3	Cell 4	Cell 5	Cell 6
MAE R_{DCIR} in mΩ	0.85	0.41	0.09	0.49	0.18	0.19
MRE R_{DCIR} in %	2.32	1.01	0.27	1.34	0.54	0.49

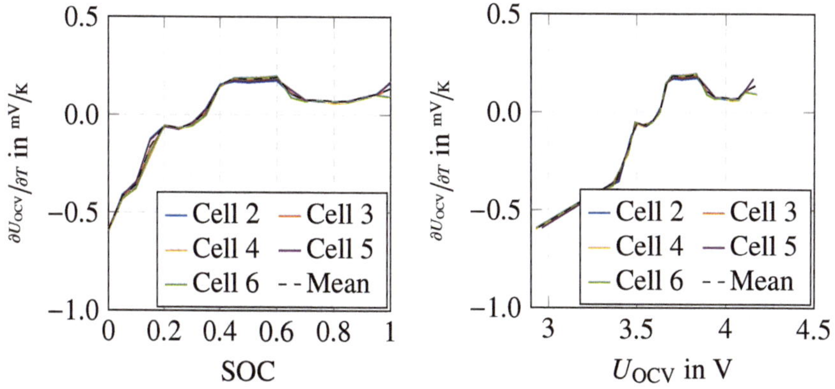

Figure C.2: Derived entropic coefficient of the five tested cells after cycle 300, along with the mean value over: a) SOC and b) OCV.

Table C.3: Mean average error (MAE) of the entropic coefficient for all five cells relative to the arithmetic mean value. The deviation range is smaller than the measurement accuracy.

Cell	2	3	4	5	6
MAE in mV/K	0.0094	0.0027	0.0105	0.0090	0.0089

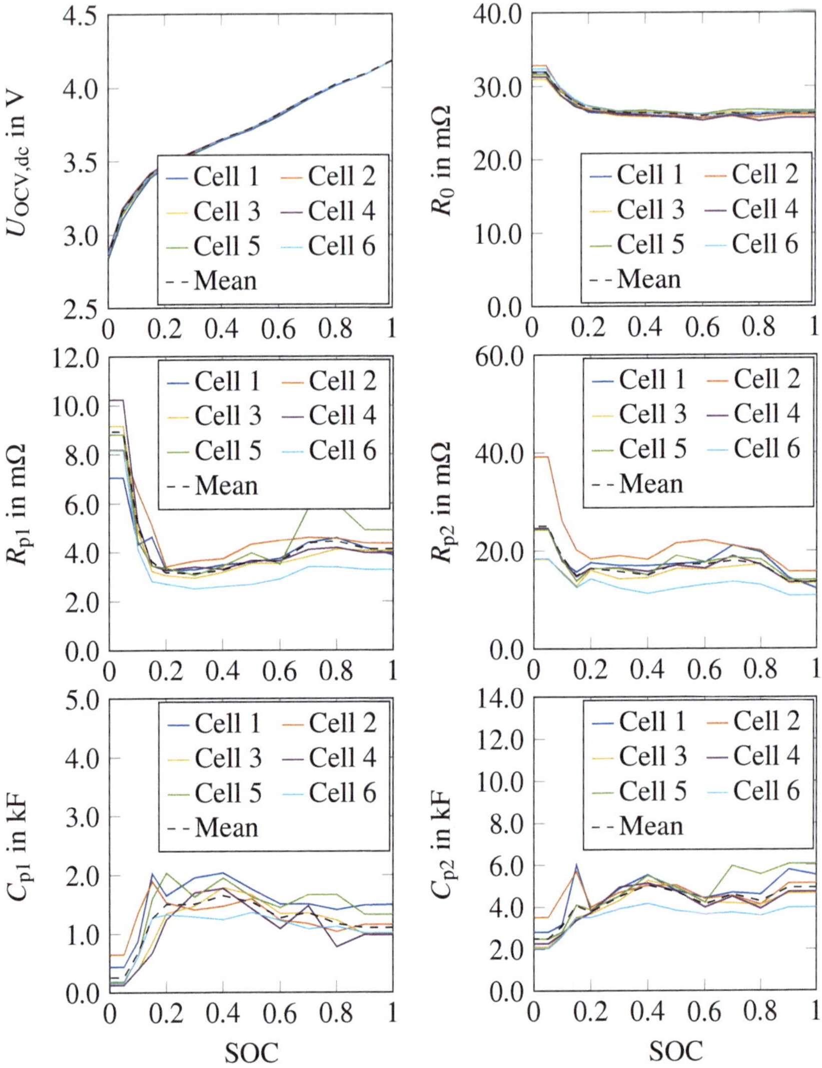

Figure C.3: Comparison of DPM parameters at SOH=100 % for the first cell (single cell) and additional five tested cells and mean value for a 0.5 C discharge pulse.

Table C.4: Mean average error (MAE) and mean relative error (MRE) of the DPM parameters of all six cells compared to the arithmetic mean value of Cells 2-6 for a discharge 0.5 C pulse.

	Cell1	Cell2	Cell3	Cell4	Cell5	Cell6
MAE U_{OCV} in mV	16.29	5	4.91	4.71	4.08	1.28
MRE U_{OCV} in %	0.49	0.15	0.15	0.15	0.13	0.04
MAE R_0 in mΩ	0.27	0.38	0.31	0.47	0.33	0.3
MRE R_0 in %	1	1.34	1.07	1.72	1.23	1.05
MAE R_{p1} in mΩ	0.51	0.62	0.19	0.33	0.49	0.79
MRE R_{p1} in %	8.95	14.55	4.32	5.53	11.51	18.53
MAE R_{p2} in mΩ	1.21	5.38	1.83	0.4	0.7	3.87
MRE R_{p2} in %	7.34	28.29	8.87	2.28	4.08	21.66
MAE C_{p1} in kF	0.31	0.23	0.13	0.21	0.23	0.13
MRE C_{p1} in %	33.85	40.32	17.76	23.69	21.3	13.67
MAE C_{p2} in kF	0.46	0.54	0.27	0.24	0.48	0.63
MRE C_{p2} in %	11.39	16.27	7.41	6.25	10.65	15.5

Table C.5: Derived mean ECM discharge parameters of Cells 2-6 at $T = 25\,^{\circ}$C.

SOC	I in A	U_{OCV} in V	R_0 in mΩ	R_{p1} in mΩ	R_{p2} in mΩ	C_{p1} in kF	C_{p2} in kF
100	0.96	4.185	26.92	3.90	14.83	1.63	5.08
	2.4	4.185	26.32	4.13	13.53	1.12	4.97
	4.8	4.185	26.32	4.13	13.53	1.12	4.97
	9.6	4.185	26.32	4.13	13.53	1.12	4.97
95	0.96	4.130	26.92	3.90	14.83	1.63	5.08
	2.4	4.130	26.32	4.13	13.53	1.12	4.97
	4.8	4.130	26.32	4.13	13.53	1.12	4.97
	9.6	4.130	26.32	4.13	13.53	1.12	4.97
90	0.96	4.091	26.92	3.90	14.83	1.63	5.08
	2.4	4.091	26.32	4.13	13.53	1.12	4.97
	4.8	4.091	26.32	4.13	13.53	1.12	4.97
	9.6	4.091	26.32	4.13	13.53	1.12	4.97
85	0.96	4.058	26.89	4.17	17.09	1.68	4.75
	2.4	4.058	26.24	4.29	15.36	1.17	4.61
	4.8	4.058	26.41	4.06	15.02	1.29	4.59
	9.6	4.058	26.41	4.06	15.02	1.29	4.59
80	0.96	4.023	26.87	4.43	19.36	1.72	4.49
	2.4	4.023	26.17	4.46	17.19	1.21	4.32
	4.8	4.023	26.49	4.00	16.51	1.47	4.28
	9.6	4.023	26.49	4.00	16.51	1.47	4.28
75	0.96	3.979	26.69	4.16	19.51	1.61	4.35
	2.4	3.979	26.24	4.41	17.56	1.30	4.49
	4.8	3.979	26.52	3.98	16.86	1.56	4.47
	9.6	3.979	26.48	3.85	16.61	1.56	4.55
70	0.96	3.928	26.52	3.89	19.66	1.49	4.20
	2.4	3.928	26.32	4.36	17.93	1.39	4.65
	4.8	3.928	26.55	3.96	17.20	1.66	4.66
	9.6	3.928	26.47	3.71	16.70	1.66	4.82
65	0.96	3.874	26.42	3.77	18.56	1.50	4.17
	2.4	3.874	26.14	4.00	17.55	1.33	4.41
	4.8	3.874	26.44	3.67	16.48	1.70	4.72
	9.6	3.874	26.35	3.41	15.33	1.70	4.95
60	0.96	3.820	26.32	3.64	17.46	1.52	4.14
	2.4	3.820	25.96	3.63	17.18	1.27	4.15
	4.8	3.820	26.33	3.38	15.76	1.76	4.78
	9.6	3.820	26.23	3.10	13.95	1.75	5.11
55	0.96	3.767	26.41	3.50	17.35	1.59	4.29
	2.4	3.767	26.11	3.64	17.24	1.41	4.47
	4.8	3.767	26.40	3.38	15.83	1.81	4.91
	9.6	3.767	26.35	3.16	14.17	1.81	5.30

50	0.96	3.720	26.50	3.37	17.23	1.68	4.44
	2.4	3.720	26.25	3.65	17.30	1.55	4.79
	4.8	3.720	26.48	3.39	15.91	1.86	5.04
	9.6	3.720	26.47	3.22	14.39	1.87	5.48
45	0.96	3.683	26.56	3.37	16.76	1.81	4.80
	2.4	3.683	26.32	3.45	16.15	1.60	4.92
	4.8	3.683	26.59	3.34	15.32	2.00	5.30
	9.6	3.683	26.55	3.13	14.07	1.93	5.65
40	0.96	3.649	26.62	3.38	16.29	1.94	5.19
	2.4	3.649	26.39	3.26	15.00	1.66	5.08
	4.8	3.649	26.70	3.28	14.74	2.14	5.59
	9.6	3.649	26.63	3.04	13.75	2.00	5.83
35	0.96	3.606	26.71	3.12	16.18	1.95	4.81
	2.4	3.606	26.42	3.19	15.37	1.59	4.79
	4.8	3.606	26.78	3.17	14.87	2.08	5.19
	9.6	3.606	26.74	3.02	14.02	1.96	5.41
30	0.96	3.561	26.80	2.86	16.08	1.96	4.43
	2.4	3.561	26.46	3.12	15.74	1.51	4.52
	4.8	3.561	26.85	3.05	15.00	2.01	4.79
	9.6	3.561	26.86	3.00	14.29	1.93	5.00
25	0.96	3.521	27.11	3.03	16.98	1.97	4.32
	2.4	3.521	26.76	3.15	16.04	1.51	4.15
	4.8	3.521	27.21	3.13	15.89	1.98	4.42
	9.6	3.521	27.23	3.01	15.84	1.85	4.42
20	0.96	3.478	27.43	3.21	17.88	1.99	4.23
	2.4	3.478	27.06	3.17	16.35	1.51	3.78
	4.8	3.478	27.56	3.21	16.77	1.94	4.10
	9.6	3.478	27.60	3.03	17.38	1.76	3.94
15	0.96	3.413	28.40	3.60	18.17	1.73	4.49
	2.4	3.413	27.88	3.62	14.89	1.32	4.21
	4.8	3.413	28.44	3.21	13.55	1.61	4.24
	9.6	3.413	28.36	3.34	13.27	1.25	4.04
10	0.96	3.292	30.08	4.38	22.20	1.15	3.27
	2.4	3.292	29.33	5.05	18.77	0.71	3.26
	4.8	3.292	30.33	4.44	17.51	0.80	3.07
	9.6	3.292	30.40	5.15	17.94	0.48	2.86
5	0.96	3.155	34.30	7.29	34.14	0.82	3.06
	2.4	3.155	31.89	8.93	25.04	0.24	2.66
	4.8	3.155	32.84	9.98	23.18	0.15	2.33
	9.6	3.155	31.42	13.61	22.90	0.06	2.14
0	0.96	2.862	34.30	7.29	34.14	0.82	3.06
	2.4	2.862	31.89	8.93	25.04	0.24	2.66
	4.8	2.862	32.84	9.98	23.18	0.15	2.33
	9.6	2.862	31.42	13.61	22.90	0.06	2.14

Table C.6: Derived mean ECM charge parameters of Cells 2-6 at $T = 25\,°C$.

SOC	I in A	U_{OCV} in V	R_0 in mΩ	R_{p1} in mΩ	R_{p2} in mΩ	C_{p1} in kF	C_{p2} in kF
100	0.96	4.186	26.52	3.49	13.46	1.34	4.72
	2.4	4.186	26.63	3.35	13.11	1.66	5.42
	4.8	4.186	26.63	3.35	13.11	1.66	5.42
	9.6	4.186	26.63	3.35	13.11	1.66	5.42
95	0.96	4.158	26.52	3.49	13.46	1.34	4.72
	2.4	4.158	26.63	3.35	13.11	1.66	5.42
	4.8	4.158	26.63	3.35	13.11	1.66	5.42
	9.6	4.158	26.63	3.35	13.11	1.66	5.42
90	0.96	4.120	26.52	3.49	13.46	1.34	4.72
	2.4	4.120	26.63	3.35	13.11	1.66	5.42
	4.8	4.120	26.63	3.35	13.11	1.66	5.42
	9.6	4.120	26.63	3.35	13.11	1.66	5.42
85	0.96	4.076	26.37	3.67	15.28	1.33	4.56
	2.4	4.076	26.44	3.47	14.45	1.57	5.05
	4.8	4.076	26.26	3.43	13.72	1.46	5.35
	9.6	4.076	26.26	3.43	13.72	1.46	5.35
80	0.96	4.028	26.23	3.85	17.11	1.31	4.44
	2.4	4.028	26.26	3.59	15.79	1.48	4.75
	4.8	4.028	25.89	3.50	14.33	1.27	5.29
	9.6	4.028	25.89	3.50	14.33	1.27	5.29
75	0.96	3.980	26.10	3.74	16.54	1.26	4.34
	2.4	3.980	26.22	3.53	15.42	1.50	4.76
	4.8	3.980	25.85	3.48	14.19	1.31	5.36
	9.6	3.980	26.04	3.36	13.23	1.53	5.77
70	0.96	3.934	25.97	3.64	15.98	1.20	4.24
	2.4	3.934	26.18	3.47	15.05	1.51	4.77
	4.8	3.934	25.80	3.46	14.04	1.34	5.43
	9.6	3.934	26.18	3.22	12.14	1.82	6.35
65	0.96	3.887	25.98	3.79	16.60	1.22	4.11
	2.4	3.887	26.11	3.54	14.96	1.41	4.46
	4.8	3.887	25.75	3.45	13.91	1.25	4.87
	9.6	3.887	26.08	3.07	12.68	1.68	5.40
60	0.96	3.838	25.99	3.93	17.21	1.24	3.99
	2.4	3.838	26.05	3.60	14.87	1.32	4.14
	4.8	3.838	25.69	3.43	13.77	1.16	4.31
	9.6	3.838	25.98	2.93	13.22	1.52	4.53
55	0.96	3.784	26.05	3.80	16.94	1.28	3.91
	2.4	3.784	26.15	3.52	15.05	1.39	4.18
	4.8	3.784	25.69	3.40	13.75	1.08	4.28
	9.6	3.784	26.06	2.96	12.72	1.53	4.76

50	0.96	3.733	26.11	3.66	16.67	1.33	3.84
	2.4	3.733	26.25	3.45	15.23	1.46	4.21
	4.8	3.733	25.68	3.37	13.72	1.00	4.25
	9.6	3.733	26.14	2.99	12.23	1.53	5.00
45	0.96	3.690	26.19	3.52	16.29	1.40	4.09
	2.4	3.690	26.33	3.33	15.05	1.54	4.43
	4.8	3.690	25.82	3.29	13.68	1.12	4.52
	9.6	3.690	26.25	2.91	12.07	1.66	5.27
40	0.96	3.657	26.27	3.38	15.91	1.47	4.35
	2.4	3.657	26.42	3.21	14.88	1.63	4.66
	4.8	3.657	25.95	3.20	13.63	1.24	4.79
	9.6	3.657	26.35	2.83	11.91	1.80	5.54
35	0.96	3.631	26.37	3.23	15.18	1.58	4.61
	2.4	3.631	26.50	3.05	14.58	1.71	4.69
	4.8	3.631	26.01	3.12	13.90	1.23	4.65
	9.6	3.631	26.45	2.81	12.62	1.90	5.32
30	0.96	3.597	26.47	3.08	14.46	1.70	4.89
	2.4	3.597	26.58	2.88	14.28	1.79	4.71
	4.8	3.597	26.06	3.04	14.17	1.22	4.52
	9.6	3.597	26.54	2.79	13.34	2.00	5.13
25	0.96	3.548	26.73	2.93	14.80	1.66	4.35
	2.4	3.548	26.83	2.73	14.45	1.79	4.33
	4.8	3.548	26.24	2.92	14.22	1.16	4.25
	9.6	3.548	26.74	2.69	13.04	2.01	4.96
20	0.96	3.497	26.98	2.78	15.15	1.63	3.84
	2.4	3.497	27.07	2.58	14.61	1.79	3.95
	4.8	3.497	26.43	2.79	14.28	1.09	3.99
	9.6	3.497	26.94	2.59	12.74	2.02	4.78
15	0.96	3.449	27.52	2.96	13.38	1.07	3.34
	2.4	3.449	27.74	2.82	13.51	1.32	3.45
	4.8	3.449	26.86	3.25	13.42	0.75	3.33
	9.6	3.449	27.37	2.71	12.78	1.44	3.71
10	0.96	3.381	28.84	3.79	14.58	0.62	2.64
	2.4	3.381	29.04	3.68	15.09	0.79	2.77
	4.8	3.381	27.83	4.35	15.64	0.48	2.65
	9.6	3.381	28.27	3.50	15.52	0.99	2.81
5	0.96	3.269	31.28	5.79	16.21	0.33	2.31
	2.4	3.269	31.11	5.45	17.02	0.39	2.30
	4.8	3.269	29.00	6.34	17.88	0.26	2.25
	9.6	3.269	29.80	5.05	17.95	0.52	2.45
0	0.96	2.862	31.28	5.79	16.21	0.33	2.31
	2.4	2.862	31.11	5.45	17.02	0.39	2.30
	4.8	2.862	29.00	6.34	17.88	0.26	2.25
	9.6	2.862	29.80	5.05	17.95	0.52	2.45

D. Modeling

Table D.1: Measured geometric parameters outer diameter d_o, inner diameter d_i and length l of the studied LIB cell.

	d_o in mm	d_i in mm	l in mm
Jelly roll	20.60	3.80	66.90
Casing	21.10	20.60	70.15
Positive terminal	20.60	-	3.00

Table D.2: Density ρ, specific heat capacity c_p and thermal conductivity k of the additional module section level materials.

Material	ρ in $^{kg}/_{m^3}$	c_p in $^{J}/_{kg\,K}$	k in $^{W}/_{m\,K}$
Aluminum [97]	2702.0	903	237.00
Copper [97]	8940.0	386	398.00
Polyurethane [117]	34.8	1407	0.03

Table D.3: Interpolated pressure independent fluid properties temperature T, density ρ, dynamic viscosity η, specific heat capacity c_p and thermal conductivity λ of EGL 50:50 [118].

T in K	ρ in $^{kg}/_{m^3}$	η in cP	c_p in $^{J}/_{kg\,K}$	k in $^{W}/_{m\,K}$
273.15	1090	9.572	3412	0.421
283.15	1085	5.571	3442	0.418
288.15	1082	4.741	3458	0.417
293.15	1079	3.912	3473	0.416
298.15	1076	3.082	3489	0.415
303.15	1073	2.607	3504	0.413
308.15	1070	2.314	3519	0.412
313.15	1067	2.021	3534	0.411

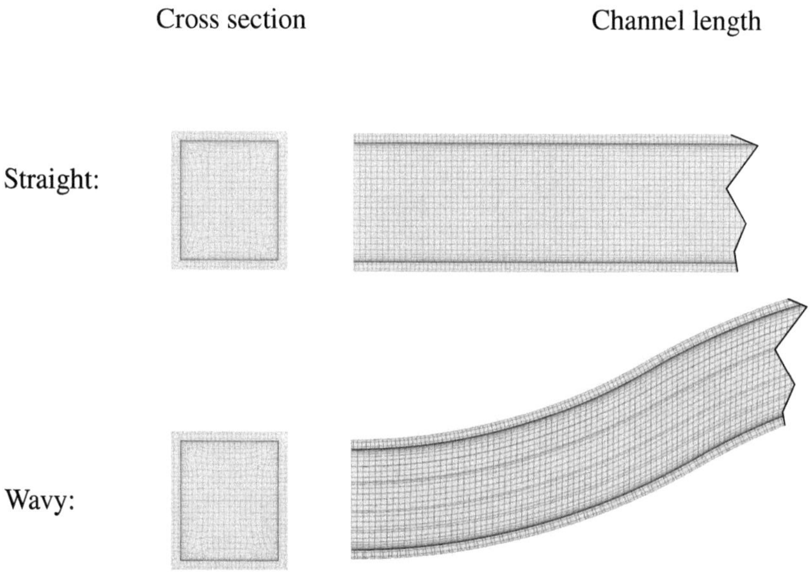

Figure D.1: Cross-section and length wise section view of the mesh for straight and wavy mini-channel heat sink.

Table D.4: Grid convergence study results for the straight rectangular mini-channel with EGL 50:50 at $Re = 187$, $T_{\mathrm{in}} = 25\,^{\circ}\mathrm{C}$ and $q = 8000\,\mathrm{W/m^2}$.

Mesh	Fine	Medium	Coarse	Fine+PL
Base size in mm	0.1	0.2	0.5	0.1
Number of prism layers	10	10	10	15
Nu_{lo} at $y^* = 0.005$	6.78	6.74	6.72	6.78
Nu_{lo} at $y^* = 0.01$	5.51	5.46	5.43	5.52
Nu_{lo} at $y^* = 0.1$	3.67	3.64	3.58	3.66
Nu_{avg} at $y^* = 0.1$	4.16	4.12	4.07	4.16
Δp in Pa	12202	12084	11837	12203

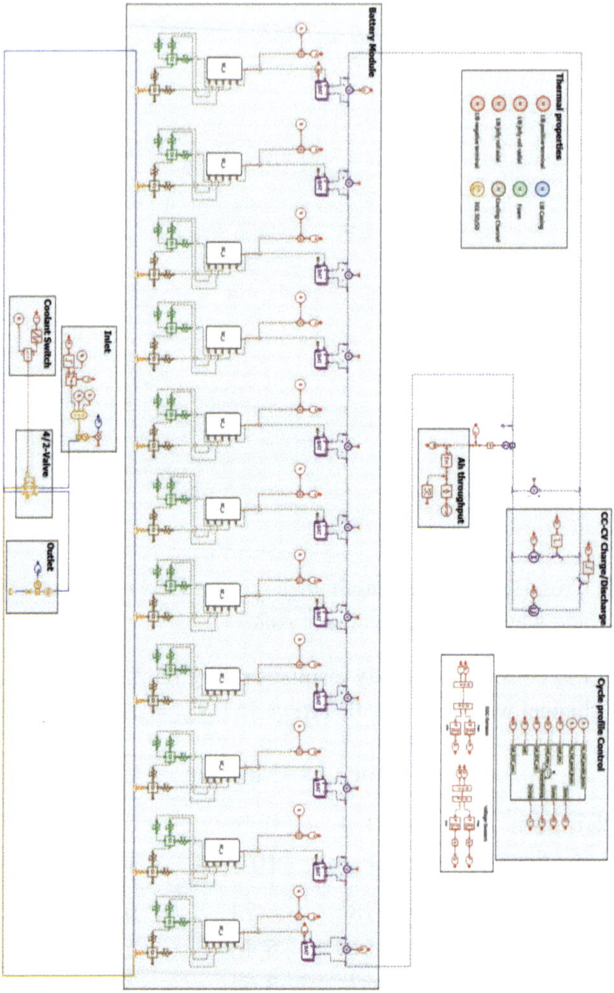

Figure D.2: Sketch of the developed Amesim model for the switchable thermal management system.

E. Validation

Table E.1: Assumptions for the derivation of the load profile based on the Tesla Model 3 Long Range LG 5L "M50F".

Parameter	Value
Battery Pack Capacity in kWh	79
Empty mass vehicle in kg	1847
Test mass vehicle in kg	1999
Mass factor	1.05
Drag coefficient	0.23
Rolling resistance coefficient	0.008
Frontal area in m^2	2.22
Powertrain efficiency	0.80
Total amount of LIB cells	4416
Cells in parallel	46
Cells in series	96
WLTP range at BOL in km	560

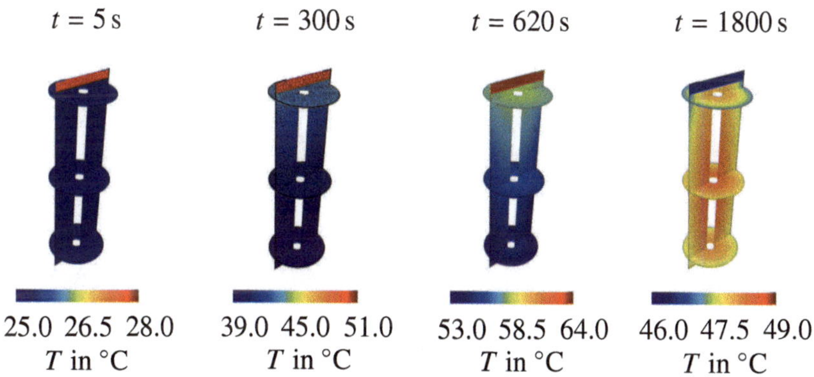

Figure E.1: LIB cell temperature during the studied fast charge scenario at different time steps. Adapted from [79].

F. Results

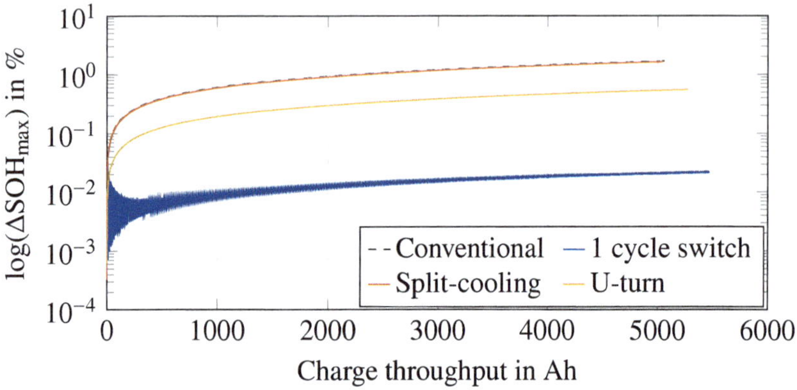

Figure F.1: Maximum SOH difference ΔSOH_{max} between minimum and maximum brick for conventional, 1-cycle switch, split-cooling and Tesla's U-turn thermal management system. Data cut off, when the first brick reaches EOL criterion.

G. Disclosure on the use of Artificial Intelligence

Declaration

During the preparation of this work, the author used the following tool and services.

Table G.1: Disclosure on the use of Artificial Intelligence.

Name of tool	Reason
ChatGPT	Polish sentences and grammar check

After using this tool, the author reviewed and edited the final content of this thesis. The author takes full responsibility for the content of this work.

MIX
Papier aus verantwortungsvollen Quellen
Paper from responsible sources
FSC® C105338

If you have any concerns about our products,
you can contact us on
ProductSafety@springernature.com

In case Publisher is established outside the EU,
the EU authorized representative is:
Springer Nature Customer Service Center GmbH
Europaplatz 3, 69115 Heidelberg, Germany

Printed by Libri Plureos GmbH
in Hamburg, Germany